CULTURE, PLACE, AND NATURE
Studies in Anthropology and Environment
K. Sivaramakrishnan, Series Editor

Centered in anthropology, the Culture, Place, and Nature series
encompasses new interdisciplinary social science research on
environmental issues, focusing on the intersection of culture, ecology,
and politics in global, national, and local contexts. Contributors to
the series view environmental knowledge and issues from the multiple
and often conflicting perspectives of various cultural systems.

Spawning Modern Fish

TRANSNATIONAL COMPARISON IN THE MAKING OF JAPANESE SALMON

Heather Anne Swanson

UNIVERSITY OF WASHINGTON PRESS

Seattle

Spawning Modern Fish was made possible in part by a Distinguished Associate Professor Fellowship Grant from the Carlsberg Foundation (CF17-0872).

Composed in Warnock Pro, typeface designed by Robert Slimbach

UNIVERSITY OF WASHINGTON PRESS
uwapress.uw.edu

LIBRARY OF CONGRESS CATALOGING-IN-PUBLICATION DATA
Names: Swanson, Heather Anne, 1979– author.
Title: Spawning modern fish : transnational comparison in the making of Japanese salmon / Heather Anne Swanson.
Description: Seattle : University of Washington Press, [2022] | Series: Culture, place, and nature | Includes bibliographical references and index.
Identifiers: LCCN 2021057680 (print) | LCCN 2021057681 (ebook) |
    ISBN 9780295750385 (hardcover) | ISBN 9780295750392 (paperback) |
    ISBN 9780295750408 (ebook)
Subjects: LCSH: Salmon fisheries—Japan—Hokkaido—History.
Classification: LCC SH350.J3 S93 2022 (print) | LCC SH350.J3 (ebook) |
    DDC 333.95/65609524—dc23/eng/20220330
LC record available at https://lccn.loc.gov/2021057680
LC ebook record available at https://lccn.loc.gov/2021057681

♾ This paper meets the requirements of ANSI/NISO Z39.48-1992 (Permanence of Paper).

For my parents, whose love, caring, and commitment first embraced me along the banks of the Columbia River and has since supported me in my travels.

CONTENTS

FOREWORD

Japanese salmon, the main subject of this creative study, are not merely a nationally identified fish stock in territorial waters of the Japanese nation. They are, as Heather Swanson shows, fish that have been altered in shape, size, and—perhaps even more fundamentally—as culturally meaningful nonhuman lives in Japan. The salmon that take Swanson traveling in Japan and to Chile played a part, she writes, in making fish and fisheries integral to the production of modern Japan. In this way, she offers a more-than-human analysis of the intertwined history of fish, fish industries, and the way Japanese identity is forged in the transformation of coast, food culture, and salmon in the coastal regions of Hokkaido in northern Japan. Food, examined here in the diverse ways salmon enters Japanese cuisine and dining, works simultaneously to produce powerful regional cultures and to illuminate how these cultures are entangled in global flows.

This is a project in multispecies ethnography, an emergent field of inquiry in environmental anthropology (see, for instance, Kirksey and Helmreich 2010 for a useful collection of essays about the contours and concerns of this field of study). Swanson learns from this line of work in her own research on salmon in Japan and Chile to examine how encounters between human endeavor and aspiration and other life-forms generate mutually transformed ecological conditions in distinct socio-spatial locations. In discussing the coproduction of human-salmon worlds in Japan in the many transnational flows that inform the constitution of Japanese food and material culture, while paying close attention to how people and salmon change together, Swanson moves beyond a narrative of management or domestication, even as such discussions remain vital to recognizing the enormous influence of human thought and action in the creation of multispecies worlds (see, for example, Cassidy and Mullin 2007).

Another significant contribution of this study is its attention to processes of nation-state and national culture formation and relating them to the making of multispecies worlds. In that sense, it is a commendable effort to bring political economy and questions of regional scales—and expanding or contracting, or even specifically aligned networks of connection across human geographies and natural landscapes—into view and to recognize their influence in constituting more-than-human ecologies. A welcome consideration of geopolitics, empire, nationalism, and environmental transformation is thus introduced into multispecies environmental anthropology, a field that at times seems removed from discussions of economy, power, and inequality.

Along the way, Swanson learns from environmental humanists as they engage with the discussion around the Anthropocene. Her sustained meditation on how new accounts of the concerns that drove preoccupation with the Anthropocene can effectively study human biological and geological transformation of the world—within still salient and, in fact, vital considerations of processes such as colonization, domination, expropriation, and the export of environmental hazards and toxicity—is central to the way she develops the analytic of comparison that organizes the whole project.

Swanson develops this view of comparison through her wide-ranging examination of specific landscapes of ecological change and human history-making within Japan and in Japanese fisheries development work in Chile. As she notes, economic development, national pride, and international commerce or industrial consulting, as well as environmental sustainability projects that follow in the wake of development and trade, are all grounded in a comparative stance. People in small communities or across nation-states evaluate themselves in relation to others who may be socially and spatially proximate or distant. This roving gaze, embodying aspirations and desires, as well as assessments of self-worth, animates the change that emerges from such evaluative actions. Swanson considers how acts, practices, and memories of comparison are central to fashioning the lifeworlds of people and salmon in Japan. In her account, Japanese salmon emerges as a food, trade good, scientific invention, and species of fish in the highly interconnected world of fish and fisheries, where salmon become a national and natural asset in modern Japan.

In a fascinating series of chapters that reveal her wide reading, sprawling engagement with locations in Japan, of course, but also with the northwestern United States and Chile, Swanson accounts for the history of development of salmon fisheries in northern Japan and a national interest in salmon across Japan as food and commodity. This takes her from earlier

Japanese awareness of northwest coast salmon in the United States to a later involvement in what appear to be suitable sites for cultivating Japanese salmon in Chile as an international development project. In the process, the study uncovers a now familiar inequality and environmental injustice created by the transnational relocation of industrial operations—in this case, fish farming—whereby greening and sustainability can reshape Japanese fisheries even as environmental degradation and pollution are exported to Chilean locations. This process is insightfully discussed in terms of shadow ecologies (see Dauvergne 1997), a concept developed to describe timber trade between Japan and Southeast Asia.

As Swanson returns, with fine-grained ethnography, to Hokkaido to consider the next round of making and remaking of Hokkaido salmon and associated industry in contemporary Japan, she also examines the debate and disagreement around belonging in the land and how that is defined by histories of association with salmon. This brings her to the stories of Ainu and Indigenous accounts of relations with salmon, which reveal the tensions of empire and nation-building around a fish and its biological alteration in changing relations with the human communities that are intensely entangled with the lives of the fish. Overall, the project spans interest and instructive findings in environmental anthropology, the political economy of food commodity chains, and social studies of science and technology. In this way, Swanson clears some new space for biological sciences, environmental humanities, and political economy of development to meet and engage in useful ways that can benefit scholarship in all these fields. This is accomplished by Swanson's close observation of the bodies of fish as she attends to the large-scale socioeconomic webs of connection in which salmon travel in altered physical form.

K. SIVARAMAKRISHNAN
YALE UNIVERSITY

ACKNOWLEDGMENTS

It is not possible to thank everyone who has helped to bring this book into being, but I would like to name a few.

Those who introduced me to Columbia River salmon worlds: Gus Fennerty, Robert Warren, and Mike Josephson.

Those who taught me to observe, question, and write in ways that shaped this project: Laura Sellers-Earl, Zaz Hollander, Andy Dolan, Andrea Kennett, Rena Lederman, Carolyn Rouse, and Will Howarth.

Those who helped me to achieve some degree of proficiency in Japanese: Setsuko Soga, Naoko Yamamoto, Sakae Fujita, Cornell University's FALCON program, SIL Sapporo Nihongo Gakko (especially Harumi Shima), the Inter-University Center for Japanese Language Studies, and the volunteers at Sapporo's Mado program.

Those who offered intellectual community, guidance, and introductions in Japan: Takami Kuwayama, Yutaka Watanabe, Masahide Kaeriyama, Noboru Ishikawa, Mohashi Gergely, Shiaki Kondo, Masahiro Koizumi, Mikine Yamazaki, Koji Yamazaki, Nozomi Aruga, and Satsuki Takahashi.

Those who actively participated in this research project: Aliaky Nagasawa, along with the many people whose words are in this book but who are referred to by pseudonyms, especially the Motozumi family, who invited me into their home.

Family and friends in Japan: the Miyoshi-Suzuki-Morichi family (especially Hiroko, Yuko, and Mariko), Yoshie Hirukawa, the Ohno family (especially Tamaki and Osamu), the Baba family (especially Yukiko), Yufuko Mochizuki, and Toshihiro and Yumiko Matsuda.

Family and friends in other places: Peter Christensen, John Law, Pierre du Plessis, Bodil Selmers, Rachel Cypher, David Pollard, Linda Oldenkamp, Matt Loftis, Kate Muslinger, Noa Vaisman, Ofer Ravid, Meredith Root-Bernstein, Katy Overstreet, and Maxine Swanson. Zac Caple, Daena Funahashi, Annika

Capelán, and Gitte Nielsen not only provided much needed encouragement, but also helped with the seemingly endless practical tasks needed to finish a book.

Those who provided intellectual guidance, collegial advice, and friendship: Marianne Lien, Donna Haraway, Andrew Mathews, Knut Nustad, Michael Hathaway, ann-elise lewallen, and Karen Hebert, along with many others. Special thanks to Alan Christy, who helped with some of the translations, in addition to other advice.

The institutions who provided funding: UC Pacific Rim Research Program (Advanced Graduate Fellowship), UCSC STEPS Institute, UCSC Center for Agroecology and Sustainable Food Systems, UC Chancellor's Dissertation Year Fellowship, the scholarship programs of the Inter-University Center for Japanese Language Studies, and the Carlsberg Foundation's Distinguished Associate Professor Fellowship program (Grant CF17-0872).

Those who enabled this project to achieve publishable form and who were patient through its many delays: Lorri Hagman, Joeth Zucco, Ben Pease, Judy Loeven, Scott Smiley, and four generous reviewers.

Lastly, there are three people who have been so important to this project—and my life—that they deserve the final spot in this list of thanks: Anna Tsing, and my parents, John and Jan Swanson.

I would also like to express my gratitude to the publishers who have granted permission to reprint sections of the following texts in this book: (with coauthor Gisli Palsson), "Down to Earth: Geosocialities and Geopolitics," in *Environmental Humanities* 8, no. 2 (2016): 149–71, courtesy of Duke University Press; "Landscapes, by Comparison: Practices of Enacting Salmon in Hokkaido, Japan," in *The World Multiple: The Politics of Knowing and Generating Entangled Worlds*, edited by Keiichi Omura, Grant Otsuki, Atsuro Morita, and Shiho Satsuka, 105–22 (London: Routledge, 2019), reproduced by permission of Taylor and Francis Group, LLC, a division of Informa PLC; "Shadow Ecologies of Conservation: Co-Production of Salmon Landscapes in Hokkaido, Japan, and Southern Chile" *Geoforum* 61 (2015): 101–10, used with permission from Elsevier.

NOTE ON ROMANIZATION

The romanization of Japanese terms follows the modified Hepburn system, except for those that have standard English expressions, such as Tokyo, Hokkaido, and Sapporo.

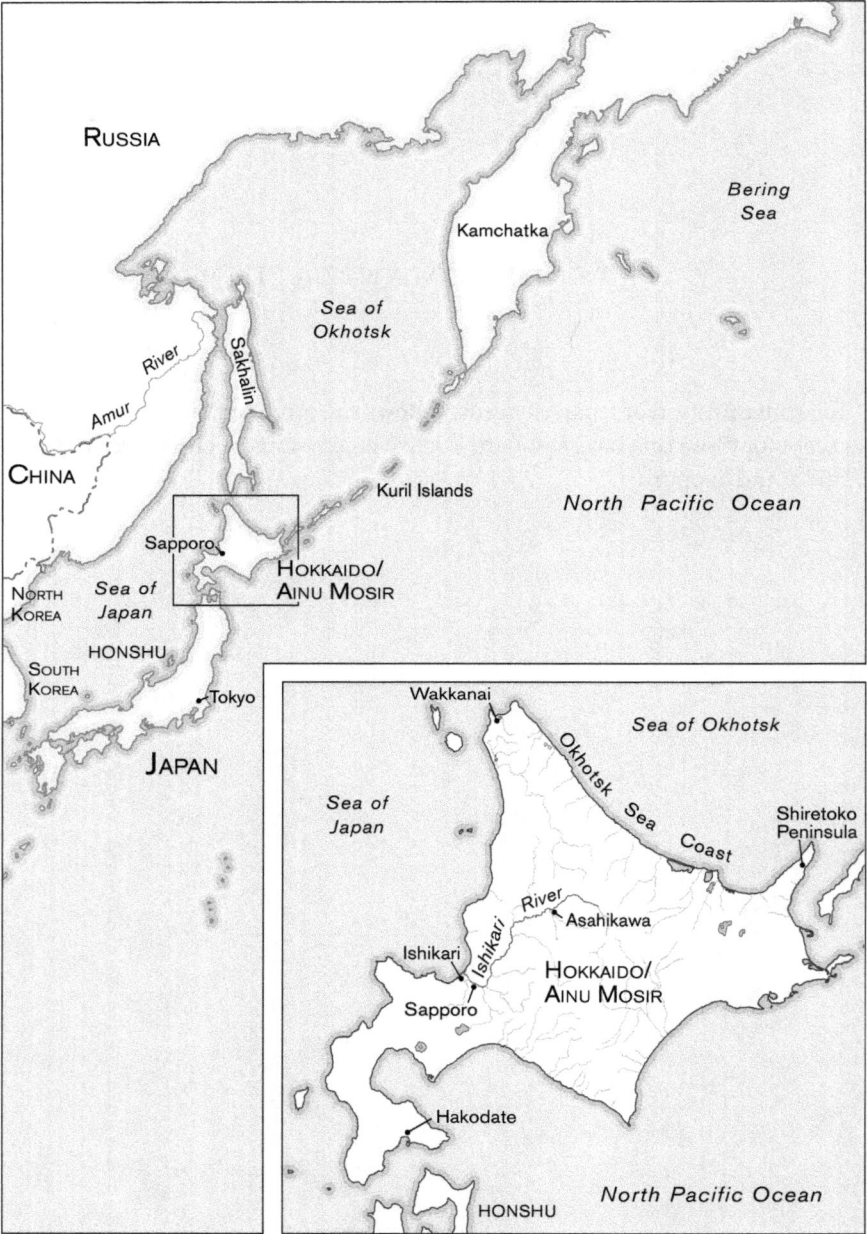

Map of Japan-Okhotsk Region, with Hokkaido detail. Map by Pease Press.

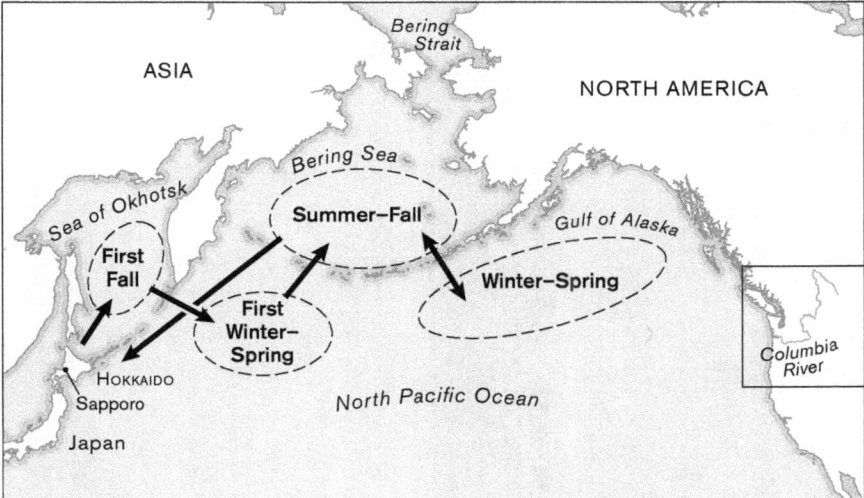

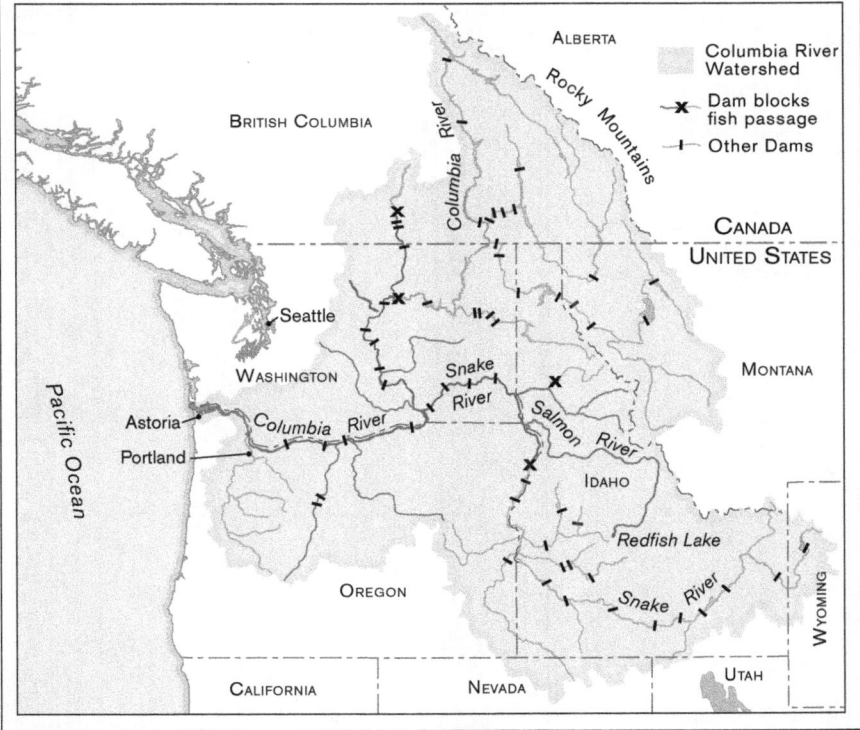

Map of North Pacific, with Columbia River detail. The top map shows the migrations and residence areas of Hokkaido chum salmon (*Oncorhynchus keta*). The ovals are the residence areas. Map by Pease Press.

SPAWNING MODERN FISH

Introduction

Material Comparisons

THIS book begins with the body of a Hokkaido chum salmon, a fish born and harvested in the coastal waters of Japan's northernmost main island as part of the nation's largest salmon fishery.[1] The salmon in the photo below had just been unloaded from a boat and placed in a metal holding crate to be sold at a dockside fish auction, shipped to China and perhaps then to a European supermarket. This book will follow Hokkaido salmon to many places, beginning from a focus on the fish themselves. As any fisheries biologist will tell you, landscape changes remake salmon bodies, as the effects of drainage, river straightening, agricultural runoff, and logging practices seep into the waterways where these fish spawn. In the case of this salmon, you are looking at a being whose life and tissues are dramatically different from what they were in the mid-nineteenth century, a result of both habitat changes and fisheries management decisions. Its body is smaller due to the cumulative effects of fishing. It spent an extra year in the ocean in comparison to its ancestors to compensate for feeding competition from other hatchery fish and for food-chain disruptions from climate change. It has returned to its spawning river earlier in the season as a result of breeding practices that have selected the earliest returning fish. And the genes of this salmon are detectably different from those of the nineteenth century, as it is the progeny of those who thrived in metal tanks and on pelleted diets rather than in streams.

This is a book about salmon, which springs from and nurtures curiosity about such changes in aquatic worlds. Yet it is also about anthropology and the growing field of the environmental humanities more generally. By working from the rapidly changing bodies of Hokkaido salmon, *Spawning Modern*

Hokkaido chum salmon (*Oncorhynchus keta*). Photo by author.

Fish asks what anthropology can contribute to interdisciplinary research on environmental issues and how the discipline might analytically benefit by further expanding its engagements with ecological assemblages. I open with a particular Hokkaido salmon because this book seeks to move beyond analytics that discuss human impacts on environments in generic terms. It aims to foreground how more-than-human relations are specific and situated, bound up in webs of political economy and relations of power. This Hokkaido salmon—whose bones, genes, and scales have been shaped by imperial projects, capitalist markets, and transnational exchange—offers a powerful example of how geopolitics matter beyond the human. When we begin to examine this fish closely, its smaller size, altered migratory timing, and adaptation to hatchery rearing show us how practices of comparative nation-making reconfigure landscapes, ecologies, and the lives of individual beings. In doing so, this book asks why scholars, conservation professionals, and others might need ethnography and history alongside things like genetic testing, fish tagging, and trap-based capture surveys to understand fish—and, by extension, multispecies relations more broadly. It presents an analytical approach that seeks to enrich descriptions of how more-than-human worlds become damaged, and in doing so, to open new questions about how they might be made more livable.

◆ ◆ ◆

Mariko Miyoshi insisted that before I left Hokkaido, Japan, we needed to go to Ishikari, a coastal town on the island's west side, for an elaborate salmon lunch. During the past year and a half, I had been researching salmon management practices in Hokkaido, the prefecture with Japan's largest seafood harvests, where salmon, the second most economically significant product after scallops, have recently had an annual value of about a half a billion US dollars (although such numbers fluctuate substantially).[2] Miyoshi-san thought it imperative that I visit Ishikari's famous salmon restaurant, a place exclusively dedicated to the preparation of this fish.[3] A spry and talkative woman in her early eighties, Miyoshi-san was an enthusiastic informal guide to Hokkaido's history. Her paternal grandfather—a farmer and veterinarian from Shikoku—had been among the first generation of settlers to colonize the island after the Japanese government officially annexed it in 1869. Miyoshi-san herself had been born in Ishikari, once home to the island's most spectacular salmon runs, and she was thrilled at the chance to take me on an outing to her birthplace.

I had been to Ishikari several times before in the course of my research, but never to dine. In the late nineteenth century, it was the site of Hokkaido's first salmon cannery and its first fish hatchery, and it continues to be home to Hokkaido's most prominent salmon processing company. The restaurant was an homage to the region's history, both in its cuisine and its decor. Stepping through its sliding door and into a low-ceilinged wooden building, I was led past a large glass display case filled with more than a dozen hair combs with tortoiseshell inlays, a pair of lacquered hair sticks, several porcelain bowls placed in the spaces between a dusty gramophone, a 1960s Nikon camera, and a small velvet-lined case set open to reveal a war medal. Miyoshi-san, her daughter, and I were seated in a private room with a view of a manicured Japanese garden. In one corner, a dark chest topped with a blue and white decorative plate sat next to a vanity cupboard with a round, European-style mirror but no legs, designed for a woman who wanted to apply makeup while sitting on the floor rather than in a chair. On the opposite wall hung a series of small black-and-white photos of boats, nets, and salmon piled on a rocky beach.

The meal's first course arrived quickly and consisted of six cold dishes, each featuring a different preparation of salmon: *kanshiobiki* (air-dried salted fish), *hiza* (pickled nose cartilage), *ikura* (roe), *izuke* (partially fermented in rice), *mefun* (salted salmon blood), and *tomoae* (salmon liver mixed with miso paste). Next came a fried salmon heart, then a grilled slice of fillet, then a pan-fried sperm sack with a side of grated daikon radish. *Ruibe* (frozen sashimi) and two pieces of deep-fried salmon wrapped in nori and served on a *shiso* leaf soon followed. The final item was Ishikari *nabe*, a local hot-pot dish made with salmon, tofu, and leek in miso broth.

The meal was at once a full-body celebration of salmon and a regionally specific performance of modern Japan as a place in deep dialogue with other locations. This transnational engagement was fundamental to the material objects and arrangements of the restaurant; one could see it in the juxtaposition of the imported gramophone and the hair combs made by Japanese artisans and in the design of the vanity cupboard that intentionally echoed European styles yet adapted them for a different mode of sitting. Nearly every object had been shaped by histories of encounter with distant places, including the war medal, which strongly echoed German designs, and even the blue-and-white porcelain, whose designs emerged through its production for European consumers who fawned over its "exotic" charm. The material culture of Japan—at the salmon restaurant and beyond—shows both how relations with other places have been so central to modern Japanese-ness and how those relations have shaped the physical form of objects. One can see histories of contact within them.

Such transnational connections have done more than shape the cultural artifacts of modern Japan. They have also made their way into less obvious material forms, such as the configurations of watersheds and the bodies of the animals and plants who inhabit them. When at the restaurant I used my chopsticks to pluck one of the last pieces of salmon from the miso *nabe* broth, I was touching the light pink flesh of a fish physically shaped by past and present encounters between Japan and other places and by the tensions of building a nation that is at once relentlessly Japanese and wholly modern in international spaces.

This, then, is a book about the making of Japanese salmon in Hokkaido—about the historical specificity of their scales, bones, and tissues. How, it asks, do processes of nation-making shape nonhuman bodies alongside human ones? Nation-making is a process of imagining community, remaking people's identities, and bringing a national culture into being through diverse processes ranging from public celebrations to acts of violence and war. But attention to salmon bodies shows us how Japanese nation-building fundamentally shapes other beings as well. It points to the ways that fish have become entangled with both state-sponsored and vernacular modes of Japanese-ness to a degree that they, too, might be productively understood as "Japanese."

Indeed, in routine fisheries parlance, salmon are often referred to with an adjective indicating the region where they were born or harvested—as, for example, Russian salmon, Alaskan salmon, or Japanese salmon. Rather than dismissing such terms as mere assertions of national ownership, this book takes them seriously as one of the starting points for its inquiries. How, it asks, are salmon pulled into projects of Japanese-ness, especially as they are enacted on Hokkaido, an island at once rich in fish and remade by Japanese settler-colonial projects?

SCALE AND SPECIFICITY

The humanities and social sciences have much to contribute to more nuanced understandings of multispecies worlds.[4] Today, it is widely accepted that humans and other beings have long co-shaped each other and that many landscapes classified as "nature" have emerged through relations with people. Yet not all human activities are compatible with lively more-than-human worlds; ecologies are suffering the effects of climate change, ocean acidification, logging, agricultural development, and urban growth. Scientists, writers, and artists grapple for terms to describe the growing scale and depth of the disruption, including the Sixth Extinction, Anthropocene, catastrophe, and crisis.[5] Although emerging out of natural science conversations, the Anthropocene in particular has raised significant debates in the humanities and social sciences. Scientists initially coined the term to emphasize that human activities have become such a strong driver of the conditions for life on earth that the planet has in effect entered a new geologic epoch, the Anthropocene, in which people constitute the most dominant world-making force (Crutzen and Stoermer 2000). As a concept, the Anthropocene has spread rapidly, generating a new sense of urgency around the mounting ecological crises caused by particular human activities—from species extinctions to radioactive contamination to the proliferation of plastic waste—especially among humanities and social science scholars who had not previously centered questions of environmental damage in their own work.

At the same time, the Anthropocene has sparked vigorous critical debates about the processes it names and thus the time period in which it began (Lewis and Maslin 2015). Some scholars have insisted on terms such as Capitalocene or Plantationocene to emphasize how particular structures and relations of power, such as capitalism or the monocrop plantation, are the driving forces of large-scale ecological harm, not the universal and undifferentiated human conjured by the word *anthropos* (Haraway 2015; Moore 2017). Such debates have focused attention on three critical processes: fifteenth-century European imperialism, extractive capitalism in the New World, and Indigenous genocide; the invention of the steam engine in 1784 and the subsequent industrialization of the nineteenth century; and the Great Acceleration, the period of rapid economic growth immediately after World War II (see McNeill and Engelke 2016). These transnational historical events are undoubtedly useful for understanding large-scale ecological transformations. Yet nestled within them is a form that has received comparatively little attention in Anthropocene debates: that of the nation-state.

Arising in nineteenth-century Europe and rapidly spreading around the globe, the nation-state was central to both industrialization and the Great Acceleration. As the nation-state coalesced nearly three hundred years subsequent to imperial capitalism, it harnessed and amplified its economic and racial logics. While this book does not engage in debates about the Anthropocene as such, it aims to speak to them indirectly by probing the role of the nation-state form and its political economies, from the nineteenth century onward, in the transformation of more-than-human worlds. Scholarship does not need any more -cenes, but if one were to characterize this book in such terms, it would be fitting to call it a critical analysis of the nation-state-ocene in an effort to highlight the importance of this structural unit to contemporary multispecies arrangements. Economic historians have written extensively about the role of nation-states in processes of industrialization, capitalization, and economic expansion (e.g., Magnusson 2009), emphasizing that nation-states have fostered growth directly through legal and financial instruments as well as indirectly through the construction of hard infrastructures, such as roads and harbors, and the establishment of softer infrastructures such as mass educational systems that prime workers for particular labor regimes (Gellner 1983). Logics of economic growth, one of the prime drivers of global environmental change, cannot be divorced from the nation-state as a unit of political ambition and power. In light of the strong role that nation-states have played in both nineteenth-century industrialization and in the Keynesian economic development of the Great Acceleration, they deserve a more central role in more-than-human scholarship as constitutive forces of environmental transformations. To be clear, this is not a call for a return to nation-state-centric analyses or histories. For anthropologists, an examination of nation-state logics is not merely a study of national policy documents; it is also an ethnographic analysis of how such logics both travel long distances and manifest in everyday life.

This kind of approach requires attention to specificity as well as to broad national and transnational trends. Analytically, it asserts the importance of describing "big" shifts and structural process, such as those of global political economy and environmental change, while also paying attention to the highly specific ways in which people in grounded places engage and shape their multispecies worlds. This book aims to undertake such multi-scalar work by foregrounding how global political and economic processes as experienced in particular places come to shape more-than-human worlds. To put it another way: How are political-economic structures lived as they change the structures of one's cells?

WHY MATERIAL HUMANITIES?

This question emerges from and speaks back to conversations in cultural anthropology, science and technology studies, and the environmental humanities. In general, the book seeks to engage central conversations in these fields in three ways:

By contributing to a humanistic scholarship that examines more-than-human worlds in their material forms. This book starts with *bodily form*, and when it describes a given act or process as "shaping salmon bodies," it means that literally, at the level of genes and phenotype. It explores how humanists might better notice the histories of social relations that shape the forms of bodies and landscapes. This focus on *embodied histories* has emerged via extended conversations with a group of scholars that has stretched across the University of California, Santa Cruz, and Aarhus University, Denmark, including Anna Tsing, Donna Haraway, Andrew Mathews, and Zac Caple.[6] Each uses slightly different terms to explore the jointly social and natural histories that adhere in bodies and worlds. Haraway (2008) asks about inheritances in the flesh as she queries whom and what she touches when she reaches out toward her dog; Tsing (2015) aims to develop "arts of noticing" the social relations that sit in the shapes of forests; Caple (2017) proposes a "critical landscape ecology" that brings landscape patterns into view; and Mathews (2018) probes how the forms of chestnut trees emerge at the intersection of political struggles, trade-borne diseases, and economic policies. This book is indebted to their conceptual work, draws on some of their terms, and aims to advance overlapping conversations (Tsing et al. 2017; Tsing, Mathews, and Bubandt 2019).

By expanding the contact zone between political economy and biology within multispecies scholarship. Fields such as political ecology and critical cultural geography have shown how state practices, regimes of ownership, and commitments to particular visions of economic development are key to understanding the making and remaking of landscapes. However, such scholarship has rarely followed political and economic conflicts fully into biological worlds.[7] How might those interested in political ecology expand their scope to explore how the processes at the center of their work take on evolutionary force, affecting the lives of more-than-human beings in addition to those of people? As we will see in the coming chapters, salmon emerge from multiple relations, and thus, a study of them requires attention to geology, hydrology, climate, and ocean conditions. But attention to biophysical characteristics alone cannot explain the evolutionary and morphological shifts in these fish; the role of political economy is too substantial to ignore. Through a range of comparative development and management practices, specific attempts to negotiate the tensions of Japanese political-economic relations make their way into the flesh and bones of salmon, altering their presents and futures. Humanists and social scientists are experts in probing the constitutive force of relations of power and the material consequences of colonial, national, and modernizing projects. Yet it is important to extend this thinking in dialogue with biological scholarship on anthropogenic change to consider how specific practices of political economy, such as those of nation-building, modify genes, bodies, and ecological configurations. In lieu of critiquing biologists for glossing complex and unequal social processes as "anthropogenic," it is essential to consider how we might better probe the ways that relations of power matter to concrete cases of organismal and ecological change. This is an act of probing the evolutionary agency of politics—not merely the political agency of nonhumans.

By describing the specificity of a natureculture assemblage. In an era of growing awareness about the entanglement of human and nonhuman lives, it is no longer a surprise that culture is shaping nature. Within the social sciences and humanities, terms such as Donna Haraway's *naturecultures*, with no space or hyphen in the word, have been particularly important in drawing attention to how human and more-than-human lives are bound up with each other (Haraway 2003). But what about the contingent, historical specificity of such naturecultures? To open up these questions, this book explicitly avoids asking about "nature" and "culture" as general categories. Instead, it asks how practices of enacting modern Japan get inside the bodies of fish. If we are to understand how cultural history, political economy, and identity shape the evolution of animals and plants, we must see environmental changes, such as the physical and genetic remaking of Japan's salmon, not as undifferentiated incarnations of global industrialization but as the product of specific landscape histories that emerge within situated transnational relations.

Engaging the Natural Sciences

Is it possible to directly engage these humanities and social science conversations while also reaching out to the natural sciences? If you are a natural scientist—especially a fisheries biologist—this book was written for you, too, even as it makes some overtures specifically to anthropologists. As indicated above, this book is deeply inspired by scientific research on salmon populations and watershed ecology. At the same time, it is committed to exploring what the humanities might contribute to biological thinking. For the past several decades, humanists and social scientists— including those working within the field of science and technology studies (STS)—have largely viewed science and scientists as objects of study rather than as allies in scholarship and world-making projects (Swanson 2017). This book moves toward the latter approach, seeking to learn about salmon together with the fisheries scientists who map their genetics, peer at the marks in their scales and ear bones, and trace how the nutrients from their carcasses make their way into the wood of the trees near their spawning streams. I consciously refuse to see biology as an epistemological other to the humanities, opting instead to see it as a discipline filled with thoughtful scholars who hold concerns that are different from, but partially overlapping with, academics located in the humanities. It is a good moment for such work, one in which the spaces of overlap are enlarging; as the humanities increasingly turn toward materiality, biology is becoming increasingly historical. Where twentieth-century biology was largely dominated by the search for universal laws to describe what were seen as ahistorical processes, contemporary biology more often views processes such as organismal development and evolution as historical and contingent. Determinism is on the wane, displaced by attention to plasticity. Many biological subfields such as ecological evolutionary developmental biology (often referred to as eco-evo-devo) now share metaphors with feminist theory and gender studies more often than they do with classical economics. In this moment of partial convergence, focusing on embodied histories seems likely to spark additional opportunities for collaboration among the humanities and natural sciences (Swanson 2017).[8]

Engaging Environmental History

While history is a relatively new site of interdisciplinary synergies between genetics and humanities fields, environmental history has spent decades developing history as a practice of multidisciplinary and multispecies research. I am intellectually indebted to a genre of environmental and frontier history that developed in large part through research on the American West that is sometimes referred to as "New Western History."[9] Beginning in the 1980s, historians such as Patricia Nelson Limerick (1988), William Cronon (1991), Richard White (1991), and Donald Worster (1985) rejected celebratory narratives of the American West as tales of a preordained triumph of civilization and national progress, instead approaching the region critically as a place of imperialism, expropriation, resource extraction, and economic expansion. Rather than fetishize the cowboy as a symbol of American freedom, they focused on structures of corporate finance, government rangeland management, industrial cronyism, and elite control. In doing so, these scholars concretely presented how powerful capitalists reoriented the region's ecologies in ways that maximized short-term returns but often left landscapes in ruins. While this genre of history stresses the importance of materiality, including particularities of weather, soils, and climate, it does so with an emphasis on contingency, not determinism, opening up other possible futures for the American West by showing how the region's violent settler colonialism, American Indian disenfranchisement, and military-industrialization were not Manifest Destiny.

Equally important in the context of this book are the spatial and temporal units of this kind of environmental history. Scholars working within this tradition have constantly taken landscapes or places as their units of analysis, posing questions about the layered more-than-human histories through which they have come into being. This spatial unit is inseparable from New Western History's temporal frames, which often explode standard historical periodization by asking how a place might be simultaneously shaped by Little Ice Age glaciation and recent land-use practices (White 1980). This mode of environmental history has itself emerged in part through the study of some of the same topics and places as those featured in this book, as several highly regarded environmental history texts, including Richard White's *Organic Machine* (1995), Joseph Taylor's *Making Salmon* (1999), and David Arnold's *Fishermen's Frontier* (2008), describe the remaking of salmon populations along the West Coast of the United States, while Brett Walker develops the analytical approaches of American environmental history in relation to the landscapes of northern Japan (2001, 2004).

Yet the impact of environmental history on anthropology has been rather muted. When New Western History hit the academic stage in the 1980s and 1990s, it did not pique the interests of contemporaneous anthropologists, who at that time were wrestling with questions of reflexivity, representation, and the politics of "writing culture" (Abu-Lughod 1991; Clifford and Marcus 1986). However, at present, the rise of more-than-human anthropology is generating more cross-pollination across these different scholarly trajectories. While this book primarily positions itself within anthropological debates, I hope that it might also be of interest to environmental historians curious about how approaches and concepts from multispecies anthropology might expand their scholarly toolboxes.

COMPARISON: A KEY PART OF MATERIAL WORLD-MAKING

For all readers, one of the book's core offerings is its attention to *comparison*, a long-debated and multiply reconfigured notion in anthropology, which also has broader relevance for understanding environmental change. Practices of comparison play a crucial role not only in the discipline but also in the embodied histories of Japanese salmon and thus also sit at the center of this book. Let us return for a moment to the restaurant and to the bites of salmon with which this introduction began. As I dined with Miyoshi-san, the very bones and genes and population structures of the Hokkaido salmon on which I chewed had been shaped by a complex web of connections that stretched to places as far-flung as a row of Oregon canneries, a southern Chilean river, and London dinner tables. Of course, Hokkaido salmon are no newcomers to relations with people. For countless generations, their lives were intertwined with those of the Ainu peoples who harvested them both for their own use and to exchange with ethnic Japanese traders (Walker 2001). But from the second half of the nineteenth century, Hokkaido salmon were pulled into a new set of projects: a series of agricultural and fisheries experiments that sought to make the island's rivers and watersheds into a model landscape for "modern Japan." Comparisons sat at the core of these efforts, in terms of both their conceptualization and their implementation. In a world dominated by Euro-American knowledges and gunboats, Japanese officials saw the development of a modern nation-state comparable and legible to those of Europe and the United States as a necessity, first for avoiding Western colonization and later for being recognized as a first-rate power on the international stage.

Anthropologists have previously illustrated how comparison-making is central to acts of nation-building, as well as to colonization. Within such contexts, people often define themselves and others by actively marking similarities and differences. Of course, comparison-making long predates nations, as people have compared their own customs, religious practices, and subsistence practices to those of others, pointing out distinctions and creating group identities. Nation-states, however, have fostered new and distinct modes of comparison. From the nineteenth century onward, the nation-state has been so naturalized as a unit of comparison that one tends to forget its relatively recent origins. It has swiftly become a taken-for-granted ground of comparison not only within geopolitics but also within social science analysis and everyday life. As other scholars have pointed out, the naturalization of the nation-state is integral to the form itself. Nation-states have sought to legitimate their rule by crafting themselves as "imagined communities" with distinct national cultures, as well as units of economic taxation and military/police power (Anderson 1983).

With these intertwined political, economic, and cultural dimensions, nation-states have become central units of comparison and comparability in a wide range of contexts and registers, from GDP metrics to the United Nations assembly to World Fair exhibits to social science analyses. Comparisons of these kinds nearly always take on a de facto mode of evaluation. They are almost never neutral and often recast colonialist tropes of racial superiority and race-based anxieties: Is one's nation-state lagging in per capita income? Are Chinese students outperforming European and American students in science and math? What can be done about the "failure" and corruption of African nation-states? Is Japan lagging behind in gender equality? Comparing well matters; it establishes geopolitical legitimacy and power. Yet the grounds of nation-state-centric comparisons have been constituted around normative ideals emergent from particular histories. The invocation of Europe and the United States explicitly and implicitly in the examples is not coincidental. While the nation-state form has often served as a tool to resist and oppose European colonial governance, it offers no simple escape from it: to be a "good" and "modern" nation-state requires deep comparative dialogue with structures that take Euro-America as a normative point of reference.[10]

Comparative dilemmas of this kind were especially important in Japan beginning in the mid-nineteenth century, when Japanese elites began to build new forms of state governance and national identity. Hokkaido was one of the places where experiments with these new modes of being Japanese were especially pronounced—and where they radically remade ecologies and landscapes. Just after the Meiji Restoration, the new Japanese state saw acts of imperial development and settler colonialism as powerful tools for constructing a legible modern nation-state, and Hokkaido became the first place where they experimented with these new genres of territorial control. Since at least 1200 CE, Japanese merchants had traded extensively with the island's Ainu peoples to obtain a share of the island's salmon harvests to supplement the much smaller harvests of northern Honshu, generally in increasingly exploitative and oppressive ways (Segawa 2007). But in 1869, the new Meiji government staked an official claim to the island, renamed it Hokkaido, and began to transform it into a landscape of Japanese frontier settlement in a new and unparalleled way, one that more substantially usurped Ainu lands and forcibly assimilated Ainu peoples.[11]

This northern land, however, was very different from the other Japanese islands; it was too cold for growing rice and was already inhabited by Indigenous peoples. Nineteenth-century Japanese government officials thus sought out overseas models for how they might turn the island into an exemplar of intensive production and modern frontier-making. The nearby Russian Far East offered an example that was climactically equivalent, but the Meiji government quickly classified Russia in official documents as a "second-rate country," enlightened but not fully civilized.[12] Hokkaido officials opted instead to focus on thinking comparatively with the American West, a place they saw as unambiguously "modern." Through this work, Japanese officials began to envision Hokkaido as a frontier where they could test and refine the most cutting-edge Euro-American ideas of the times, including forms of scientific agriculture and modern fisheries management. With the help of invited American experts, Japanese officials crafted a suite of transnational comparisons that would radically reconfigure Hokkaido's landscapes, importing new breeds of livestock and new kinds of seeds from the United States, constructing dairy farms with American-style barns and silos, and planting rows of potatoes and corn with the same sod-breaking plows used to turn under the Kansas prairie.

Such comparisons were material projects; officials drained Hokkaido's wetlands, converting them into fields for industrial agriculture at the same time that they channelized and dammed the rivers to protect farms and provide irrigation water. These so-called river improvement projects—coupled with increasing agricultural runoff, forest clearing, and chemical use—damaged the spawning grounds for the island's salmon. Simultaneously, these fish were also directly enrolled in governmental modernization schemes. Inspired by the lucrative salmon canning operations along the US West Coast, Hokkaido ramped up salmon fishing in the 1870s, built their own canneries based on American models, and began exporting tinned fish to Europe on a large scale as they sought to develop comparable products and modes of industrialization.

As high harvest levels and habitat degradation decimated salmon numbers, Hokkaido officials studied and compared the latest in scientific fish propagation techniques from Europe and the United States, quickly establishing a system of salmon hatcheries, as the fish could no longer spawn effectively in Hokkaido's channelized rivers. At these facilities, technicians bred salmon by hand, mixing together strains of fish from geographically distant rivers (including some from the United States) in ways that transformed the genetic structures of Hokkaido's salmon populations. Such practices led to a sharp decline in river-spawning salmon numbers and further altered the ecosystems in which they had once been a keystone species.

Today, comparisons and the changes that they have inscribed in landscapes are central to fisheries management in Hokkaido. When I conducted the anthropological and oral history fieldwork that undergirds this book, I encountered ongoing comparisons that at once inherited and differed from those of the nineteenth century: a university fisheries school modeled after an American land grant college, salmon scientists who tried to distinguish their theories of sustainability through comparisons with Canadians, and members of a salmon fishing cooperative who had designed their business practices in comparison with models from Russia.[13] Everywhere I went in Hokkaido, people cited relations between their own fisheries practices and those of people in Norway, France, New Zealand, and Chile. In northern Japan, no one I met in the field of salmon management did anything—from hatchery fish rearing to post-harvest processing to scientific research design—without constantly referencing geographically distant sites. These ricocheting sets of comparisons continually create cross-border movements, including introductions of new species, exchanges of currency, transfers of scientific technology, and exports of products, which, in turn, remake the identities of Hokkaido's people, the uses of the island's terrain, and the genes of its fish. While such quotidian practices may seem far removed from the comparisons of nation-state geopolitics, one of my arguments is that everyday comparisons at the interface of social practice and fish flesh are fundamentally intertwined with geopolitical structures.

I want to emphasize that this book is not a comparative study; it does not analyze salmon fisheries in different places by comparing them. Rather, it is an ethnographic exploration of how people make comparisons and how those comparisons affect material worlds. The following chapters examine the ongoing practices of comparison of fisherpeople, scientists, government officials, Indigenous peoples, and environmental activists in relation to historical practices of comparison-making in Japan, especially those embedded in efforts to make Japan a powerful nation-state comparable to those of Europe and the United States. These histories are significant because past comparisons linger and present comparisons happen in worlds shaped by those made before them. People do not get to craft comparisons de novo but find themselves entangled, even caught, in the living legacies of earlier comparisons.

As we dive into Hokkaido's salmon worlds, we shall see that people's practices of comparison shape nonhuman worlds along with social categories. Efforts to perform modern Japanese fisheries are clearly intertwined, for example, with fish harvesters' intense desires to craft themselves as cosmopolitan businesspeople comparable to other major players in the international seafood trade. Yet the need for comparability and the comparisons they compel are also among the forces that drive the watershed and fisheries management practices that shape the bodies and populations of Hokkaido's salmon. Comparisons do not stay in people's minds but instead seep out into the world. They inspire actions that change material arrangements and forms. Consider once again the Hokkaido salmon at the beginning of this chapter. As part of the quest to modernize Hokkaido and to have comparable forms of industrial fisheries and agriculture, Japanese agencies and cooperatives from the late nineteenth century onward have operated large-scale salmon hatcheries that extract eggs and sperm from adult fish, fertilize and hatch the eggs, then rear and release young fish into rivers, from where they migrate to the ocean and back on their own. These hatchery practices have almost certainly altered this fish's genes, as workers have stirred together the gametes of salmon from different rivers around Hokkaido that would otherwise be unlikely to spawn with each other, at the same time that the facilities' metal tanks and feeding practices have exerted new evolutionary selection pressures. As salmon like this one have come to start their lives in hatcheries rather than rivers, they have become different beings.

Attention to the salmon that have been so important to Hokkaido's history reveals how people's comparative practices are landscape-making forces, not just ways of knowing. Overall, through the case of Hokkaido salmon, this book argues that the ways people make comparisons in a world permeated by nation-state logics constitute a substantial but often overlooked evolutionary force and driver of ecological change. Salmon are often

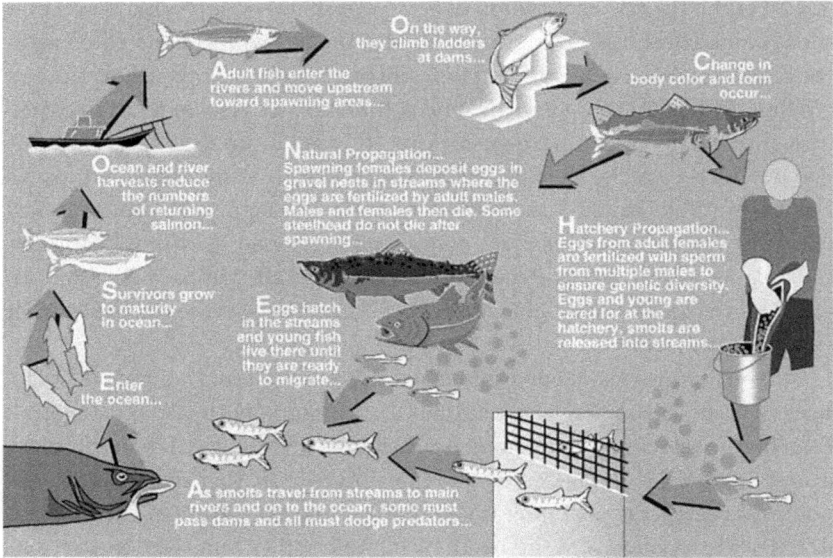

Salmon life-cycle diagram, showing both stream- and hatchery-based modes of reproduction. Courtesy of the US Army Corps of Engineers.

caught in nets, and it is established scientific knowledge that the specifications of fishing equipment and the practices of harvest exert selective pressures that alter fish genes, bodies, behavior, numbers, and more; yet salmon also get caught in—and get made by—structures of comparison.

HOW DOES THIS BOOK USE THE TERM *COMPARISON*?

The way this book uses the term *comparison* is likely to be disorientating for some readers. In general, comparison is considered to be a cognitive act, an estimation or measure of difference or similarity. If the length of two straws is compared, comparison is not seen as occurring in the straws but in the mind of the person who compares them. In contrast, this book argues that comparisons are not exclusively mental acts but rather material practices in which mind and body are fundamentally intertwined. Until recent years, anthropologists have typically thought of comparisons in one of two ways: either as analytics, as frames laid on top of already existing worlds, or as modes of thinking that shape our ethnographic descriptions and interpretations. Like a handful of other texts, such as Timothy Choy's

Ecologies of Comparison (2011) and Shiho Satsuka's *Nature in Translation* (2015), this book instead explores comparisons as world-making practices. This approach is partially inspired by similar assertions within the field of science and technology studies; STS scholars commonly use the term *knowledge practices* rather than *knowledges* to signal that knowledge-making is performative, that is, that it comes into being through embodied acts, institutional arrangements, and more-than-human relations rather than within the confines of human brains (Law 2008; Mol 2003). With a resonant sensibility, this book focuses on comparative practices, with the following propositions about the materiality of comparisons:

> **Comparisons have material effects and accrete within material objects.** When a Japanese consumer compares imported and domestic salmon at a supermarket and decides to purchase the fish labeled "Hokkaido," when a Chilean biologist compares the temperature of a Patagonian river to one in Hokkaido and determines that it might be possible to transplant fish from one side of the Pacific to the other, or when an Ainu leader makes an appeal for fisheries rights modeled on that of an American Indian group, comparisons reconfigure Hokkaido's human and nonhuman livelihoods in a physical way. As assertions of value, possibility, and rights, comparisons compel actions that then remain in the material forms they shape.

> **When I write about comparisons in landscapes or in fish flesh, this is not a loose metaphor.** It draws from an established tradition in material culture studies of probing how knowledges and concepts become embedded in forms. One example of this scholarship is anthropologist Alfred Gell's work (1996) on animal traps. For Gell, a trap is a materialization of its maker's analysis of a particular animal's worlds. As Gell writes, "Once the trap is in being, the hunter's skill and knowledge are *truly located in the trap, in objectified form*, otherwise the trap would not work" (27, my emphasis). I extend such thinking to propose that knowledges and concepts are found not only inside human-made technologies but also inside landscapes and bodies that are remade by specific human projects.

> **My approach to the accretion of comparisons is also indebted to work on Japanese linguistics.** Japanese is a language that has been profoundly shaped by multiple borrowings—of kanji characters from

China and loanwords drawn from Portuguese traders, German diplomats, and English-language television. A single sentence often includes words drawn from different time periods with different histories of contact. While the way that language accumulates histories within it is not unique to Japanese, Japanese marks some terms as "Western" in a special way. Typically, Japanese words made through contact with Western languages are written in a block-style script called katakana, rather than in either kanji or the more cursive hiragana. Each term written in katakana is at once Japanese and not-Japanese; katakana situates words and concepts within Japanese worlds while simultaneously signaling a link to the West, as it is conceptualized and enacted in Japanese contexts. In this way, comparisons of West and East are built into the structure of contemporary Japanese; comparison sits *inside* the language itself. Comparisons come to be embodied in material forms in a similar way. A Hokkaido fish hatchery can be seen as a form of katakana in the world, a material entity that emerges within the comparisons and juxtapositions between Japan and the West and that holds the histories of those comparisons in its structure.[14]

Comparison itself is a material act, but comparisons are never predetermined by material stuff. Consider again the straws mentioned at the beginning of this introduction. When one compares two straws, one rarely uses one's mind alone; one also use one's hands. One sometimes brings the two straws together from different places, places them alongside each other, and perhaps nudges their ends so that they align in a certain way. Such bodily engagement in comparison-making is not trivial. In the case of Hokkaido fisheries, Meiji era officials and twenty-first-century fisherpeople alike often physically traveled to other places to hone their comparative sensibilities, to encounter modes of canning and textures of salmon fillets not only with their own eyes but also their own hands. Comparisons create cross-border movements, such as the introductions of new species, exchanges of currency, transfers of scientific technology, and exports of products. Yet movements of materials also spark new comparisons. For example, when farm salmon from Chile entered Japanese fish markets, they prompted a new set of comparisons and a reconceptualization of Japanese-produced fish. It is important to note that comparisons are shaped by happenstance as well as by plan. Hokkaido fisherpeople did not set out to compare the fish they harvested to new imports; they were drawn into a comparison that they did not initiate.

Fish make comparisons, but this book focuses on the comparisons of people. Salmon are agential beings who analyze their world, and they indeed make comparisons as they move through their lives—between different kinds of prey, as they select a meal, as they decide between two rivers with different smells, and as they select a place to spawn. In short, salmon make comparisons that operate via their own logics, and these practices also accumulate in their bodies. Noticing this puts the effects of human actions on salmon into perspective: people shape salmon, but *they do not make them.* Salmon have countless relationships with other beings and entities—with rocks, currents, caddisflies, and krill, to name only a handful. Salmon entered the global scene long before tensions around nationhood, capitalist economies, or geopolitical maneuvering. According to recent archeological and genetic estimates, Pacific salmon predate humans by more than fifteen million years.[15] These deep time histories remain in their bodies in a big way, and they remind us that while relations with people play an increasing role in the lives of salmon and the waters they inhabit, humans are not the only actors.

That said, I reserve the term *comparison* for the types of comparative practices and challenges with which salmon become entangled but in which they do not engage—those that invoke concepts such as the West, modernity, and nation-states. Salmon and their worlds are transformed by such comparisons, but salmon *knowledge practices* are enacted in relation to other kinds of concepts and entities. Furthermore, while intended to raise questions of broad interest, the following chapters are not about human comparison in a generic sense. They are about people who compare from and with Japan. Comparison has taken on a specific and special force in enactments of modern Japan— in a particular historical period (since the mid-nineteenth century) and in relation to particular political and economic structures insepa- rable from the nation-state form. Thus, the analyses of comparison elaborated here refer to and are emergent from this empirical context; they aim to be useful for thinking about other modes of comparison but not to be directly transportable to differently situated comparative practices.

ANALYTICAL APPROACH: WHY COMPARISON AS AN ETHNOGRAPHIC OBJECT?

What are the benefits, one might ask, of stretching *comparison* and *comparison-making* in ways that may seem slightly awkward? Wouldn't established terms and concepts such as *connections, flows,* or *cultural borrowing* accomplish similar work? In choosing to center comparison instead of one of these other concepts, this book takes up broader conversations about who gets to be an analyst and whose analyses gets to count as such.[16] As an ethnographic object, comparison blurs the lines between analysis and the world. In fields such as anthropology and history, when one uses the term *comparison,* listeners typically assume that one is speaking of a scholarly act, of an analytical attempt to think across two or more cases. Yet scholars are not the only ones who do analytical work; everyone undertakes analysis in their daily lives.

Indeed, for the Hokkaido fishing industry professionals with whom I began my research, practices of comparison-making are intentional onto-epistemological acts that are explicitly discussed and cultivated (see chapter 5). Comparison-making, for them, is at the heart of being what they variously call *kindaiteki* (modern), *kokusaiteki* (international), and *shinkashita* (advanced), concepts at the core of their efforts to improve their fisheries and cultivate themselves. They not only demonstrated but also overtly explained comparison as a technique for crafting oneself and one's fisheries through consciously learning about and skillfully negotiating worlds of multiple practices and standards. *Cultural borrowing* or other common academic phrases do not fit with the ways Hokkaido fishing industry professionals consistently articulated themselves as resourceful people who actively analyze, navigate, and intervene via comparison-making. On the contrary, they often emphasized their own innovation and creativity, insisting that they were not just "borrowing" but variously using, juxtaposing, and contrasting.

While they had honed and reflected upon their comparative practice well before I arrived on the scene, the geometries of comparison that Hokkaido fishing industry professionals described to me were also shaped by our relational encounters. As a white woman who grew up in a salmon fishing town in Oregon, along the West Coast of the United States, my very being invoked comparisons from the moment I arrived in Hokkaido, as most of the people I met posed large numbers of queries about my hometown salmon (along with other general questions about the United States) as they interpolated me into their ongoing comparison-making, with the phrase *to kurabete* (in comparison to) as a regular part of our interactions. Because many of our conversations came to revolve around comparisons between Hokkaido and the Columbia River region where I grew up, I lean into these ethnographically emergent comparisons, even as most salmon fisheries professionals would describe Hokkaido's twenty-first-century fisheries as more comparable to those of Alaska's chum salmon industry. This approach also holds true to a key feature of comparison, as the Hokkaido fishing industry professionals described and enacted it: one does not merely compare sites that seem "naturally" comparable; one also makes comparisons across sites that seem radically different.

This book focuses on comparison-making because it draws on and builds from the analytical work of these fishing professionals. This is part of its commitment to a mode of grounded anthropological practice that insists that one's analytical categories should emerge through fieldwork itself. In contrast to the natural sciences, where once in the field, researchers aim to implement a predesigned methodology as faithfully as conditions allow, good anthropological research is seen as dependent on an openness to having one's research question, one's analytical categories, and even one's most fundamental assumptions about the world upended in the midst of fieldwork.

Yet *comparison*, as I use it within this book, is more my concept than theirs. While I am inspired by the analytical insights of my interlocutors, I am not translating them. Their notion of comparison defined it in primarily epistemological terms—as a mode of thinking—even as they acknowledged its material effects, in contrast to my analyses of comparison as thoroughly material. Their comparisons also frequently deployed conceptual juxtapositions of progress and modernity alongside backwardness with a less critical approach than mine, as I situate these ideas in relation to histories of the making of the Japanese nation-state and its settler-colonization of Hokkaido. Perhaps most importantly, while I concur with their analyses of their own comparative practices as heightened and more carefully honed than those of many other people, I nonetheless emphasize continuities across time and space that they do not, as seeing their practices as both an intentional and unusual achievement and as one that gestures toward more widespread comparative dilemmas and acts—some linked to the particular contexts of Japan, but many reaching well beyond. Overall, these analytical approaches came into being as my comparisons and curiosities intersected with those of others who have interests in Hokkaido salmon.

CHAPTER ONE

Situating Comparisons

From the Columbia River to Modern Japan

I BEGIN again with a story about myself as a comparer. My first plan for this research project focused on practices of salmon management in the Columbia River basin, located in the northwestern United States and southwestern Canada. I had grown up in a small town near the river's mouth, and I had long been passionately interested in the braiding of human and fish lives within the basin. Before the mid-twentieth century, the Columbia was among the world's richest salmon-bearing watersheds, and even after decades of serious declines in fish numbers, salmon were still at the core of the region's identity and economy for many Indigenous and settler communities. Salmon were in local school curricula, in public artwork, at the community maritime museum, and, frequently, on my dinner plate.

When I began studying Pacific salmon, I thought I knew quite a bit about these fish. I was raised in a white wooden house built by an early twentieth-century salmon cannery administrator, and I had watched from my parents' bedroom window as low-gunwaled gillnet boats set drifts. I took three years of salmon biology coursework at my local high school and worked at the school's on-site fish hatchery. Later, I spent several years working at an organization that focused on salmon restoration. Based on these experiences, I thought I had a sense of the basic analytical categories and themes that would matter for an anthropological study of Pacific salmon worlds—and comparison was not among them. The units of salmon management, as I had learned them, were watersheds, fishing zones, farm fields, and irrigation district boundaries. The debates were about how to allocate fish stocks among commercial, tribal, and recreational fishers and how to best protect "wild" salmon. Although political machinations in Washington, DC, often impacted fisheries management, salmon issues were consistently considered a regional concern. In my years studying and working in salmon management in Oregon and Washington State, the people I encountered—be they hatchery workers, fisheries scientists, government officials, or environmental activists—all mapped out salmon worlds that were linked to a specific watershed (i.e., to the Columbia River basin) or, at the largest, to a salmon ecoregion that stretched from coastal California to Alaska (Woody, Wolf, and Zuckerman 2003). In their everyday practices of salmon management, people along the Columbia River did not often think globally or draw comparisons to far-off sites. In my countless interactions in Columbia River hatcheries, on fishing docks, at dams, and at meetings, I do not recall anyone even mentioning the existence of Japanese salmon.

I first learned about these salmon many years later when, as a PhD student, I opened a copy of *The Atlas of Pacific Salmon* (Augerot et al. 2005), a collection of geographic information system (GIS) maps that depicted salmon populations around the entire Pacific Rim, stretching in a nearly contiguous arc from California to Japan. I was dumbstruck to learn that there were salmon in Asia—in Kamchatka, Siberia, Hokkaido, and even Honshu. The wilds of eastern Russia seemed roughly comparable to Alaska and thus somehow comprehensible as salmon spaces. But salmon in Japan? It was a phenomenon that had never crossed my mind. At once, my curiosity was doubly piqued. First I wanted to know what Japanese salmon worlds were like. Then I wanted to know how it had been possible for me to not know about them.

Motivated by this growing awareness of my ignorance, I decided to reformulate my research to focus on salmon-human relations in northern Japan rather than in the Columbia River basin. Hokkaido seemed the place to start; while there are salmon hatcheries and commercial fishing in Honshu, the industry is much larger up north, with total adult salmon numbers about four times larger in Hokkaido than in Honshu when I was beginning field research in the late 2010s (Hokkaido National Fisheries Research Institute 2020).[1] After I arrived in Hokkaido, however, I realized that the geographies of Japanese salmon management were far larger and more cosmopolitan than those I knew from the Columbia River and that I thus needed to reconceptualize my research. While fisheries professionals in the Columbia River rarely invoked comparisons with places beyond the bounds of the North American West Coast, in Japan, globe-spanning comparisons were one of the most prominent features of the salmon industry. Most of the fisherpeople, scientists, and Indigenous activists in Hokkaido knew of the Columbia River and the town from which I came, and some had even been there. They repeatedly pointed out the bits and pieces of Columbia River salmon worlds that had been drawn into their own—the nineteenth-century Hokkaido salmon canning label that used one from the Columbia River as its model, the Ishikari fish trap based on a Columbia River design, the Hokkaido wild salmon policies that had been directly shaped in dialogue with those of the United States. The comparisons that they showed me, however, were not limited to the Columbia River; instead, they reached out to Parisian fish markets, Chilean rivers, Alaskan management strategies, European supermarkets, and American restaurant chains.

Comparison emerged as the key theme of my research because it was inescapable; analogic thinking was both a key practice for the people I met and something they showed me as materially sedimented into Hokkaido's worlds. When I arrived in Hokkaido, I was already thinking comparatively, in the common anthropological sense of working analytically across the salmon worlds I knew from the Columbia River and those I was about to encounter in Japan. But the ubiquitous comparisons of the people I met in Hokkaido forced me to consider comparisons as far more than analytical or methodological tools. Such encounters pushed me to ask how people's practices of comparison might also be studied as ethnographic objects and material world-making practices in ways that at once build on and expand existing anthropological conversations about comparison.

ANTHROPOLOGICAL COMPARISONS

Cross-cultural comparison has always been a major methodological and theoretical concern in anthropology. It is heralded as a core contribution of the discipline, an essential humanist act that allows us to draw parallels between others' lives and our own and a practice that can denaturalize taken-for-granted assumptions of "how the world is." But at the same time, cross-cultural comparison has also been critiqued as a colonialist endeavor, one that has been used to create developmental hierarchies and racial typologies. Anthropologists are caught in the dilemmas of comparison; we are deeply wary of echoes of nineteenth-century comparative methods, but we continue to depend on comparison as one of our most important techniques. Indeed, the very notion of "culture" does not exist separate from practices of comparison. Yet even as they have enthusiastically compared, anthropologists have also recognized the fundamental incomparability of different ways of being. As the noted anthropologist E. E. Evans-Pritchard allegedly said, "There is only one method in social anthropology, the comparative method—and that is impossible" (Needham 1975, 356).

Anthropologists negotiate practices of comparison while also struggling with questions of power and politics. To compare inevitably positions someone's categories as the grounds of comparison, that is, as the analytic framework within which the comparison unfolds. Anthropology has increasingly grappled with the question of *whose* categories take on that role. In anthropology, the idea that so-called emic categories, or those of one's fieldwork interlocutors, should fundamentally shape one's analysis, questions, and interpretations has long played a substantial role in disciplinary practice. Yet scholars have become increasingly uncomfortable that Euro-American analytical categories, both within and beyond anthropology, remain too dominant. Overall, since the widespread critique of cultures as bounded entities, many anthropologists have moved away from explicit engagement with comparison, instead working through concepts such as connections, flows, and -scapes (Appadurai 1990). Might it now be useful for comparison to once more take on a more central role in disciplinary debates? If so, on what scholarly resources might a new mode of comparative anthropology draw?

In the last three decades, the interface between anthropology and postcolonial theory has been a particularly rich site for probing the categories through which cross-cultural comparisons are made. But the comparisons that have been the critical focus of well-known postcolonial theory texts—such as Edward Said's *Orientalism* (1978) to name but one—have primarily been Euro-American colonial ones, that is, those of the West, with a focus on how its comparisons create structures of power and lasting inequalities. What, then, of the comparative practices of the people who are not Euro-American? How might we better attend to the world-making effects of their comparisons?

Eduardo Viveiros de Castro, an anthropologist working in Brazil, has recently called for scholars to do just this—to take others' modes of comparison more seriously. Addressing the problem of how anthropological comparisons too often render difference only in their own terms, typically those of the West, Viveiros de Castro asks, "How can we restore the analogies traced by Amazonian peoples within the terms of our own analogies? What happens to our comparisons when we compare them with indigenous comparisons?" (2004, 4). My attention to the role of comparison in Japanese fisheries management responds to Viveiros de Castro's invitation to be curious about others' comparisons. At the same time, the historical and ethnographic stories I tell cast comparison in a different light. For Viveiros de Castro, Amazonians and Western anthropologists present two very separate, and indeed inverse, modes of comparison; Amazonians locate difference in material form and worlds, while Western anthropologists tend to assume a singular physical reality while emphasizing differences in "culture"—in perception, belief, and interpretation. Viveiros de Castro's emphasis on the contrasts between these two modes of thinking is an essential analytical move, one that engages with Latin American Indigenous thinkers and activists for whom assertions of alterity are a central part of colonial resistance.

In Japan, however, comparisons and comparative politics take on a different valence. In the case of Hokkaido salmon fisheries, there is no singular mode of Japanese comparison but rather sets of comparative practices that have come into being through historical encounters.[2] To more fully explore this historical emergence, it is useful to bring Viveiros de Castro's call into dialogue with another strand of anthropological work, that of Ann Stoler and Benedict Anderson, who have focused specifically on how comparisons come into being within colonial practices. By tracing the comparative practices of nineteenth-century white Europeans in Indonesia, Stoler has argued that European modes of comparison are formed *within* the projects of colonial administration, not prior to them. In her work on modes of racialized and sexualized governance, Stoler prompts us to think about how nineteenth-century colonial comparisons were not fully created in a European "core" and then exported to the "peripheries" but rather were made in encounters within the colonies themselves (Stoler 2001). For this reason, Stoler queries comparison-making in practice by attending to the specific biographies and trajectories of the nineteenth-century colonial officials. It was their routes and travels, she shows, that allowed them to develop particular practices of comparison. "Agents of empire," she writes, "were themselves rarely stationary. They moved between posts in Africa and Asia, schooled their children in international Swiss boarding schools, read avidly about other colonials, visited colonial expositions in Paris and Provence, came together in colonial hill stations around the globe, and had a passion for international congresses where their racial taxonomies were honed and their commonsense categories were exchanged" (853). Stoler draws our attention to how such comparative practices coalesced into more durable structures of comparison: "Category making produced cross-colonial equivalencies that allowed for international conferences and convinced their participants—doctors, lawyers, policy makers, and reformers—that they were in the same conversation, if not always talking about the same thing" (863).

While Stoler's work focuses primarily on the comparisons of European colonial bureaucrats, her historical and biographical approaches can be extended to explore how the comparative practices of non-Europeans similarly emerge within their travels and encounters. How might non-Europeans' modes of comparison also shift over time as their worlds are changed by acts of comparison-making, both their own and those of others? The work of Benedict Anderson offers additional insights here. His book *The Spectre of Comparisons* opens with a story from Jose Rizal's novel *Noli me tangere* that illustrates how colonies are haunted by comparisons with their so-called metropoles. Set in the 1880s, Rizal's story tells of Ibarra, a mestizo man who has just returned to Manila after extensive travel in Europe. When Ibarra moves through the colonial city, he discovers that he can now only see its landscapes in comparison with those of the center. Manila's municipal botanical gardens, he realizes, are forever shadowed by their "sister gardens" in Europe. Ibarra finds himself caught in the comparisons of the colonial predicament; he can "no longer matter-of-factly experience [the gardens] but sees them simultaneously close up and from afar." Rizal terms this "incurable doubled vision" the *demonio de las comparaciones*, the devil or specter of comparisons (Anderson 1998, 2).

But at the same time that Anderson clearly shows how colonial comparisons force those in the colonies into the position of the "copy," he also demonstrates how comparisons contain subaltern possibilities. In the introduction of *Under Three Flags*, Anderson offers the example of Isabelo de los Reyes, a late nineteenth-century Filipino folklorist and nationalist, who harnessed colonial comparisons to challenge the European domination with which they were entangled. Drawing on the allegedly universal science of European folkloristics and writing in Spanish, Isabelo cleverly talked back to his colonizers by using their comparisons as well as their language. Isabelo routinely placed the customs of Filipino groups alongside those of Spanish communities. By depicting Filipino and Spanish folk traditions as comparable, Isabelo sought to stake a broader claim of equivalence that would undermine colonial projects and bolster Filipino nationalism (Anderson 2005, 13–19).

As Anderson has demonstrated in much of his work, nation-making is always shot through with comparative practices. It requires "imagining community," fostering a "we" and defining its boundaries by making comparisons with constitutive outsides (Anderson 1983). But while nation-making is always a comparative process, nation-making in the midst of colonial comparisons requires double work. In centers of European colonial power, national folklorists played similarly important roles in developing imagined communities. But while European folklorists could work to conjure a relatively singular audience to which they wrote, Isabelo always had to speak to two. He had to use comparative folklore both to create a "national brother-hood" among Filipinos—a category that itself did not yet exist—and to make the Philippines a legitimate and legible nation in the eyes of European colonial powers. "If in Europe folklorists wrote mostly for their *paisanos*, to show them their common and authentic origins," Anderson explains, "Isabelo wrote mostly for the early globalizing world he found himself within—to show how Ilocanos and other *indios* were fully able and eager to enter that world, on a basis of equality and autonomous contribution" (Anderson 2005, 22).[3] In a world dominated by colonial logics, nation-making outside the metropole required more than consolidation; it required the extra work of making one's nation comparable to the core European nation-states by which it was inevitably haunted.

Yet unlike Indonesia or the Philippines, Japan has never been directly colonized. Instead, like many states on the margins of Europe, it blurs the line between colonizer and colonized. On one hand, people in Japan have been caught in unequal relations with Euro-America and in comparative predicaments that resemble those of colonial relations. But Japan, too, has occupied the position of colonizer. Beginning with Hokkaido and Okinawa, then Taiwan, Manchuria, Korea, and southeast Asia, Japanese officials undertook their own imperial projects—including the enactment of their own colonial comparisons. Furthermore, after World War II, the Japanese state has continued economic efforts and international development proj-ects imbued with colonial sentiments. Thus, instead of simply "provincial-izing comparison" and seeing it from unequivocal colonies, this book examines it from a place where multiple genres of comparative practices have been in play as Japan negotiates the challenges of being modern and non-Western, a colonizer and a nation caught up in Euro-American colonial frameworks.

COMPARING FROM JAPAN

Everywhere I went in Japan, the spectral presence of something called "the West" seemed to linger. When I checked into Japanese hotels, I was typi-cally asked if I wanted a "Japanese" (*washiki*) or "Western" (*yōshiki*) style breakfast, and when I entered a public restroom, I had to choose between stalls designated as containing either a "Japanese" or "Western" style toilet. When a Japanese friend contemplated her upcoming wedding, she debated at length whether she wanted to go "Japanese" style, wearing a kimono and holding the event at a shrine, or "Western" style, with a white wedding dress in a chapel. Japan-West distinctions were embedded in the rhythms of quotidian life, a form of so-called emic classification that has been analyzed by other scholars, including anthropologist Harumi Befu (1984). From where might these ever-present comparisons have come? To offer a broad historical answer to this question drifts uncomfortably close to stereotypes and monolithic representations of "the Japanese." Yet some sense of overarching geopolitical histories and the binary conceptual categories commonly used in everyday life in Japan is necessary for understanding heterodox comparisons-in-action.

From the outset of its self-identification, Japan has been constituted by comparisons. The islands—as feudal alliance—grew up as a comparative margin in the greater Chinese imperial domain, inheriting its written script and civilizational arts. Yet feudal Japan's ambitious elites explicitly altered them, drawing difference into political relations. Until the nineteenth century, Japanese scholars made countless comparisons with China as they struggled to define an identity at once connected to and distinct from the mainland. But in the context of such comparisons, boundaries were rather fluid. With identity linked more to differences in manners, customs, and the style of one's poetry than to a notion of cultural essence, people could slip easily between categories. Although there was a vague sense of a "Japan," feudal domains—not the "nation"—were the important grounds for identity-making. Indeed, before the nineteenth century, the word *kuni* (country) was not often used, and when it was, it "more often referred to the local region or domain than to Japan as a whole" (Morris-Suzuki 1998b, 13; see also Roberts 2002 on identity in Tokugawa Japan). Although the Tokugawa state (1603–1867) held varying levels of influence over the islands of the North Pacific, its borders were uncertain, and in outlying areas, tenuous ties to the central government were the norm.

In the seventeenth and eighteenth centuries, Japanese intellectuals began to map the world—and their relations to it—more extensively. Drawing on Confucianist models, they depicted the world in terms of concentric rings of foreignness, envisioning a geographical gradient from intimately familiar to utterly exotic. As historian Tessa Morris-Suzuki has shown, Tokugawa intellectuals, like many mainland Chinese scholars, saw difference primarily through spatial comparisons, with *i* (barbarian qualities) increasing as one moved farther away from the *ka* (the settled center) (1998b, 15). Such modes of comparisons were far from benign in either China or Japan. As we will see, they had devastating effects on Ainu communities and the island now most widely referred to as Hokkaido. Yet they constituted a mode of governance, economy, colonial rule, and cultural thought that differed dramatically from the forms that would soon come to dominate governmental practices, as well as public culture.

Beginning in the mid-nineteenth century, Japanese elites began to radically alter their comparative practices. For the previous two centuries, foreign exclusion policies had seriously circumscribed contacts with Europeans and kept trade relations under tight shogunal control. But by the nineteenth century, the growing number of British, French, American, and Russian vessels plying Asian waters made the foreign exclusion policy and the political control it provided seem increasingly untenable. From their writings, it is clear that Japanese elites were well aware of the Opium Wars and the mounting power of Europeans to subjugate China. After watching Britain force Chinese ports to open to trade, build a colony on Hong Kong, and back the Qing dynasty into signing what are now known as the "unequal treaties," Japanese elites worried about their own future. It seemed that an arrangement of power was beginning to take shape in East Asia, one in which the United States and Europe were likely to become ever more aggressive. Japanese officials and intellectuals began to fear that if they did not do something quickly, they would become a colony of the West.

The arrival of Perry's Black Ships in 1853 upended both intellectual thought and everyday life across much of the archipelago. This assertive military visit from a powerful American fleet catalyzed unrest among political elites and contributed to the Meiji Restoration, which ousted the shogunate and returned power to the emperor. However, it created more than a political regime change; it also created a shift in comparative practices. By the nineteenth century, Euro-Americans had begun relying on a mode of comparison that brought temporality into understandings of difference, such that otherness was reconfigured as backwardness and Euro-American lifeways were cast as "development" and "progress." From the late eighteenth century onward, European and American social theorists, including Condorcet, Comte, Spencer, Morgan, and Tylor, participated in the elaboration of such ideas via vigorous discussions of sociocultural evolution that positioned different groups of people along a continuum ranging from the most "primitive" to the most "civilized" (Fabian [1983] 2014). Crucially, these authors defined civilization in highly Eurocentric terms, linking it not only to Enlightenment science, high arts, formal schooling, and economic industrialization but also to emergent governance structures, such as the nation-state.

Although there was much internal disagreement, most of Japan's powerful nineteenth-century elites felt that the island's best hope for avoiding

Western colonization lay in modern nation-state formation—in making a different kind of Japan. Soon, most Japanese elites adopted ambitious plans for *bunmeikaika*, namely civilization and enlightenment, diving into the project of creating a nation in dialogue—in comparison—with those emerging in the United States and Europe. During the Meiji period (1868–1912), Japanese elites adopted a Prussian-style constitution, built an English-style navy, studied French army tactics, and instituted an American-style education system (B. Walker 2005, 129). Meiji era Japanese intellectuals and government officials worked as hard as they could to build a Japan that would be legible to existing Euro-American nations, that could occupy the same categorical level as those of Europe. Unconsolidated early nineteenth-century Japan seemed to Euro-Americans to be a terrain ripe for colonization; as a real first-rate nation, "Japan" would no longer be a target. To join the ranks of the "civilized" instead of the "backward," Japan had to become comparable to the West rather than to other Asian countries. In his famous essay "Datsu-a ron" (Leaving Asia), nineteenth-century author, translator, and educator Fukuzawa Yukichi argued that Japan needed to escape from Asia and enter the West:

> Once the wind of Western civilization blows to the East, every
> blade of grass and every tree in the East follow what the
> Western wind brings. . . . The spread of civilization is like the
> measles. . . . In my view, these two countries [China and Korea]
> cannot survive as independent nations with the onslaught of
> Western civilization to the East . . . We do not have time to wait
> for the enlightenment of our neighbors so that we can work
> together toward the development of Asia. It is better for us to
> leave the ranks of Asian nations and cast our lot with civilized
> nations of the West. (Lu [1885] 1996, 351–53)

Leaving Asia and becoming civilized entailed not only military and economic infrastructural projects (such as railroads and factories) but also efforts to create new daily habits and "modern" sensibilities. Things like meat eating, pocket watches, and umbrellas changed the rhythms, tastes, and aesthetics of everyday life in urban Japanese contexts, often as part of government-led initiatives yet also extending far beyond them. Because the changes were so material, resistance to Meiji enlightenment efforts often explicitly centered on *things*, such as gas streetlights and kerosene home lamps (Steele 2007, 59, 65). Anti-Western critics denounced them not only as symbols of enlightenment efforts but also as entities with rippling material effects, expressing concerns about how such objects might disrupt domestic economies, bind people to new regimes of purchasing, create import dependencies, produce new safety hazards such as a greater risk of fire, and—specifically in the case of lamps—alter bodies through eyestrain (Steele 2007).

Yet whether one was for or against Meiji era civilization initiatives, the categories of public debates took on a nearly singular form. As historian Carol Gluck describes it, "The Meiji frame was almost always the *kokumin kokka*, the nation-state, the national people, national progress, or its lack" (Gluck 1997, 12). Within this frame, the binary between East and West became one of the primary analytics for conceptualizing both geopolitical relations and Japanese identity—"the metaphorical coin used to debate the nature and extent of change in every corner of Meiji experience" (Gluck 1997, 13; see also Racel 2011, 71). The increasing use of East-West comparisons was part and parcel of new geopolitical arrangements and imaginaries into which Japanese elites were thrust by European and American military incursions but in which they were also engaged participants.

The writings of Fukuzawa, the previously cited nineteenth-century proponent of Meiji enlightenment efforts, gesture to the fact that the East-West binary was actively constructed within Japanese intellectual work and not only via labors of Europeans. In his *Outline of a Theory of Civilization*, Fukuzawa noted that he "equate[s] the terms 'Europe' and 'the West.' Although Europe and America differ geographically, the latter's civilization derives from Europe, and so I feel justified in using the general term 'European Civilization.' The same holds true in the case of the term 'western civilization'" (Fukuzawa 1973, quoted in Racel 2011, 83). Across otherwise substantial political and philosophical differences, Meiji intellectuals routinely conceptualized the same binary civilizational geographies, producing a pervasive comparative sensibility that shaped everyday material worlds alongside elite discourse. As Gluck has noted:

> Since acquiring civilization entailed Euro-Americanization (*ōbeika*), the "West" embodies the standard of modernity, which in turn posed the challenges of defining the "East" along a new axis of identity. The juxtaposition was palpable in such things as lamps and haircuts, powerful in institutions like parliaments and extraterritoriality, enticing in challenges to create new forms of individual subjectivity or the novel (*shosetsu*)—all relentlessly pitted against some essential Japaneseness that had itself to be improvised on the fly. (1997, 13)

It is important to note the condensations common in these conversations: civilization (*bunmeika*), Westernization (*ōbeika/seiōka*), and modernization (*kindaika*) were often used interchangeably from the nineteenth century onward, covering "roughly the same semantic domain" and linked to a shared set of comparative practices (Befu 1984, 71).

Initially, efforts to craft and inhabit such binaries appeared to pull Japan closer to the economic, military, and cultural parity with the West that such binaries constructed. As Japan embraced Western-centric forms of modernization, it began to work its way toward the top echelons of international hierarchies. Drawing on physical and institutional infrastructure from the Tokugawa period—including standardized weights and measures, an integrated road system, a wealthy merchant class, systems of credit, and extensive intraregional trade—Japanese elites were rapidly able to build both a strong economy and a powerful military. Japan's victory over Russia in the 1904–5 Russo-Japanese War displayed the success of such endeavors. The conflict marked the first time that a "non-Western" country had defeated a "Western" one, and both Euro-Americans and Japanese took note of the significance of the occasion. Just after the war, a Japanese author, in an English-language article published in the *New York Times*, wrote:

> To rise in a bound from the rank of "yellow monkey" to the
> position of a great power is certainly a most prodigious feat; yet
> this is, in a sense, what Japan has accomplished. Only yesterday
> she was regarded, at least by the Russians, as a "yellow monkey"
> with a thin veneer of civilization; to-day all nations look upon
> her as one of the world's greatest powers. (Kawakami 1906)

After this buoyant beginning to the twentieth century, Japanese intellectuals believed that the primitive/civilized continuum offered a mechanism through which they could claim the mantle of "world power." They may have been caught up in the West's culturally specific mode of comparison, but it appeared that they could achieve military, economic, and cultural parity within such frameworks.

In the early twentieth century, Japan continued to change in the midst of imperial aspirations. In the run-up to World War II, Japanese intellectuals yearned to do more than just work their way up the West's ladder. Some sincerely sought to make more just and non-Western-centric worlds through the construction of new Asian alliances; others sought to maintain, but invert, existing hierarchical structures. As historian John Dower explains, "In the modern world, [as] Japanese researchers repeatedly observed, racism, nationalism, and capitalist expansion had become inextricably intertwined. The Greater East Asia Co-Prosperity Sphere, as they described it in the abstract, would break this pattern by creating an autarkic community governed by reciprocity and harmonious interdependence." In practice, their colonial policies were "so structured economically and politically as to ensure that the relationships of superior and inferior would be perpetuated indefinitely" (Dower 1986, 266). Ultimately, Japanese governmental elites ended up flipping models of hierarchy—putting Japan at the top—rather than reconfiguring them. Keeping models of civilized/backward in place, they endeavored to replace Western pretenses of universalism with Japanese ones.

As Japanese soldiers took over increasingly large stretches of Asian territory, Japan itself changed—not only its physical boundaries but also its approaches to identity and belonging. During Taisho (1912–26) and early Showa era colonialism, the category of "Japanese" increasingly yoked together nation, culture, and ethnicity into a single unit, such that blood and nation were made isomorphic. One famous example of this nation-building scholarship is philosopher Watsuji Tetsurō's 1935 book *Fūdo*, in which he claimed that Japan's four-season climate and environmental features made the Japanese people distinct, uniquely balanced, and superior to other peoples (Watsuji [1935] 1988). Yet even frameworks that espoused Japanese superiority and incomparability were entangled in deep conversations with Europe. Watsuji developed his work, for example, in critical dialogue with that of Heidegger (Befu 1996). More generally, as historian John Dower shows in his book *War without Mercy*, "The affirmation of Japanese supremacy reflected Western intellectual influences as well as Western pressures" (1986, 265). Japanese intellectuals, he shows, drew extensively on German ideas of *Volk*, blood purity, and social Darwinism. A 1943 document written by Furuya Yoshio, a medical doctor who held a position in the Japanese government, illustrates the use of ideas that echo those used in European, including Nazi, formulations of nationalism: "No nation in this part of the Orient can stand comparison with Japan in point of racial virility and organizational ability. The racial vigor of Japan is the most potent factor that has enabled it to attain its present distinguished position in the polity of nations" (quoted in Dower 1986, 276).

After defeat in World War II, many Japanese intellectuals worried that in the process of competing with the West, they had reaffirmed the power of the West to make the rules. As early as 1948, Takeuchi Yoshimi, a Japanese scholar and prominent postwar intellectual, argued that it was imperative that the Japanese realize that there was no way to "overcome modernity"

(Takeuchi 2005)—not industrialization nor foreign exclusion nor a Greater East Asian Co-Prosperity Sphere. Japan's pathology, he stated, lay in its failure to recognize its inability to be free from the West. From his view, Japan's engagements with Western modernity had been flawed from their beginnings because Meiji era elites did not recognize that they were caught in a double bind. The Japanese, Takeuchi argues, have repeatedly failed to see that they cannot escape a world shaped by Western dominance even if they escape formal colonization. Modern comparisons, he explains, are nonoptional and, no matter how made, offer no respite from a Western-oriented world; no amount of maneuvering will lead to freedom from them. To underscore his own unavoidable intellectual entanglement with Euro-America, Takeuchi (2005) drew on Hegel's master-slave dialectic to explain how the Japanese, no matter what they do, are forced into the position of slave vis-à-vis the West.

Rebuilding Japan after the war became yet another project of making Japan differently—a project of capitalist expansion rather than military imperialism in which Japanese intellectuals dreamed of global hegemony through economic success. In the 1980s and early 1990s, as Japanese investors purchased American landmarks, including Rockefeller Center, Pebble Beach golf course, and Radio City Music Hall, it really did seem that Japan was upending—economically, at least—Euro-American dominance. But even at the height of Japan's transnational economic strength, Japanese intellectuals continued to struggle with Euro-American comparisons that characterized them as derivative. Japanese people were frequently depicted as uncreative technicians rather than as inventors; from electronic goods to pop music, their production—material and cultural—was widely depicted as imitative rather than original, reiterating established, racialized stereotypes.

Since the Japanese economic collapse in the mid-1990s, similar comparative predicaments remain present in the archipelago, including within scholarly contexts. As Japanese anthropologists have pointed out, they continue to inhabit unequal relations of academic power, which require that they know and engage Euro-American scholarship, while Euro-Americans (even those who conduct fieldwork in Japan) are free to ignore Japanese anthropological theory. If Japanese scholars allow themselves to be interested in questions different from those of Euro-American disciplinary peers, their work is often illegible to major English language journals and thus is limited to circulation within Japan. If they want their research to participate in the valorized space of international scholarship, Japanese scholars can end up caught in a catch-22; they "must conform to the dominant discourse at the center in order to be recognized," but when they do, their work is seen as unoriginal. "Conformity to the center may be derided as imitative, whereas nonconformity will likely result in dismissals of their work for being incomprehensible" (Kuwayama 2004, 39, 40; see also Asquith 1996, 2000).

The experiences of Japanese intellectuals over the past century and a half make it clear that there is no easy way out of Western-centric comparisons. Attention to such dilemmas does not absolve either the Japanese state or elites of responsibility for acts of Japanese colonial aggression, which are not any better or more legitimate than those of Europe or the United States (Kondo and Swanson 2020). Their violences cannot be justified as products of Western-generated dilemmas. Furthermore, Japanese concepts related to history, ethnicity, and difference are not inherently good because they emerge from more-than-Western traditions.

Unequal dialogues with Euro-American categories and comparisons are mandatory for people in Japan, but despite their inequalities, they are indeed dialogues, not monologues. At the same time that they are structured by nineteenth- and twentieth-century histories, they are not entirely determined by them, thus producing unexpected comparisons whose particularities warrant attention. Consider a Hokkaido citizens' study group, organized by a local NGO, in which I participated in 2009–10. The aim of the group was to develop future visions for the island that might challenge its patterns of resource extraction. For one session, participants were asked to prepare short presentations in which we compared Hokkaido to some other place as a technique for imagining alternate and more sustainable futures for the region. The vast majority of participants selected European Nordic countries—Finland, Norway, and Sweden—as key reference points for envisioning good governance, eco-friendly lifeways, gender equality, and recognition of Indigenous rights.[4] On one hand, such comparisons show the continued role that "the West" continues to hold within the geographies through which Hokkaido's futures are iteratively conceptualized. Yet these were also different comparisons with different content, which indeed sought to compare creatively and, in opposition to previous comparisons, with the aim of imagining a more just and less extractive Hokkaido. As I engage the people and salmon at the heart of this book, I try to follow the imbrications, reverberations, and contradictions of their multiple yet omnipresent comparisons—while allowing them to come up against and transform my own.

CHAPTER TWO

Landscapes, by Comparison

Hokkaido and the American West

A FEW months after he relocated to Hokkaido in 1907, the poet Ishikawa Takuboku was stirred to pen a short essay about his encounters with this new Japanese land. In the midst of domestic turbulence in Honshu, Ishikawa had traveled first to Hakodate to serve as a substitute elementary school teacher, but he soon moved on to Sapporo after the school burned down in a fire. By that autumn, he had made his way to the western port city of Otaru, where he took a job writing for one of the town's newspapers (Pulvers 2015). Despite his own misfortune and financial insecurity, he saw Hokkaido as place with a bright future. In his text, *First Sight of Otaru* (Hajimete mitaru Otaru), Ishikawa dwelled on Hokkaido's possibility and vigor:

> The spirit of the settler and the taste of the frontier endow people
> with unexpected might. Think about it—since Europe emerged
> from the deep slumber of the dark ages, a myriad of brave
> adventurers have set their sights on manly adventure in America,
> Africa, Australia and much of our Asian region. Think too how to
> this day, what was once called Yezo Island, now the island of
> Hokkaido, has pulled in countless adventurers from the main-
> land. Our Hokkaido is the land of freedom, thrown open for us
> Japanese. The children of freedom across the country, acting with
> spirit and bravery, have doubtless been stirred by that untamed
> land stretching out as it does like a continent. (Ishikawa 1967)[1]

For Ishikawa and others, Hokkaido was a vast tabula rasa, despite its Ainu Indigenous community, deep histories of trade relations with Honshu, and fast-growing commercial centers, such as Otaru. Entangled in easily recognizable tropes, it quickly became a space of virgin territory awaiting virile adventurers, a place whose bounty extended from its rocky mountain spires to its expansive seas:

> By the mountains of white clouds and setting sun, where not a single human step has been planted since the dawn of the world. By the hinterland of the great verdant forests. By the great plains, expanses of desert reticent of rural Russia. And by the limitless oceans, frothing white and swarming with fish. (Ishikawa 1967)[2]

Ishikawa's romantic prose is merely one example of the ways that Japanese officials, writers, and Hokkaido-bound migrants repeatedly framed the island as a frontier (*shinkaichi* or *furontia*) and a colonial project (*shokuminchi*) (Mason 2012a). Through the invocation of such terms, Hokkaido was nearly always thought of and experienced in relation to other places awash in similar expansionary imaginaries, practices of Indigenous disenfranchisement, and resource extraction. Yet Hokkaido was a specifically *Japanese* frontier rather than a wholly generic one. What role did comparison-making play in the development of Hokkaido as place with a complex pattern of similarities and difference, as an unambiguous "frontier" with a distinctly Japanese sensibility? In Hokkaido, the frontier was not a mere abstraction. Instead, the island's frontier framing quickly led to many concrete comparisons between it and other highly specific spots, and the specificity of those comparisons came to have significant impacts on the fish swarming in Ishikawa's endless and frothing white seas.

Hokkaido, as a place name, has always marked a comparative project. While this large island north of Honshu has physically existed for thousands of years—since the submersion of the land bridge that connected it with Russia—Hokkaido itself has a shorter history. Prior to nineteenth-century Meiji modernization, the island was known as a part of Ezo, a name that carries a meaning of "barbarian lands."[3] Ezo and its Ainu residents were firmly entangled with Japanese trade networks, but Ezo was not considered a part of Japan proper. In the frenzy of post-Restoration nation-making and increasing fears of Russian incursion, Meiji modernizers changed the name of these northern lands to *Hokkaido*, a word that means "north sea route" or "north sea district," and initiated efforts to incorporate the area into the territory of Japan through colonization and development. The change in name marked an important conceptual shift. While Ezo was generally seen as outside Japan-as-such, Hokkaido was to be "Japan's frontier," a critical site for practices of nation-making. The new Meiji state was explicit that its designs for the region marked a shift. "Today's Hokkaido is not yesterday's Ezo," declared one government document (Mason 2005, 2).[4]

Hokkaido marked a project clearly distinct from that of Ezo, one rooted in new kinds of comparative practices.[5] When Japanese officials compared their new nation to those of Europe and North America, they felt that they needed their own colonies in order to claim their place as a first-rate global power. In the nineteenth century, being internationally recognized as "civilized" was closely tied to a regime's ability to claim its "ability to transform an uncivilized people" (Dudden 2005, 3). For the Meiji state, Hokkaido, with its Indigenous Ainu people, was an ideal site to enact the kind of civilizing drama that would demonstrate Japan's potential to become the "Great Britain of the East" (Kublin 1959, 76).[6] Officials hoped to do so by enacting Hokkaido as Japan's American West, a place to demonstrate national vigor by domesticating "wild" people and "wild" landscapes. European colonial imaginaries merged with classic Western frontier fantasies, producing powerful visions of a place where the oxymoronic platitude of "peaceful conquest" could reign supreme (Mason 2012b, 39–42). Like their American counterparts, Meiji officials simultaneously described the colonization of Hokkaido as "peaceful pursuits" and as "industrial warfare and conquest" (Nitobe, quoted in Mason 2012b, 39–40).

Getting such frontier narratives "right" profoundly mattered to nineteenth-century Japanese elites. They were not content simply to settle Hokkaido and extract its resources; they wanted to do so in internationally legible ways. Making Hokkaido into a frontier was essential to making its colonization comparable to that of Euro-American nations. The desire to create a comparable colonialism is especially evident in the work of Nitobe Inazō, a Japanese diplomat and politician who attended college in Hokkaido.[7] In 1893, Nitobe wrote a pamphlet in English in which he explicitly framed the colonization of Hokkaido using language that echoed that of nineteenth-century Western colonialism:

> The northern islands of Japan, vaguely called Yezo, were for
> centuries a terra incognita among the people: all that was told
> about, and unfortunately most readily accepted by them was
> that the region was the abode of a barbarian folk known as the
> Ainu, and that it was a dreary waste of snow and ice, altogether
> unfit for inhabitation by a race of higher culture. To Yezo, then,
> at once the northern frontier of the Empire and a land endowed

with magnificent natural resources as yet untouched by human hand, the new Imperial Government wisely began to extend its fostering care. (Nitobe 1893, 1–2)

But as important as language is, enactments of "the frontier" are never done by narrative alone. They are also always material practices of landscape-making. Just as Japanese officials sought to make internationally legible narratives of frontier colonialism, they also aimed to create physical landscapes that would appear undeniably colonized in the eyes of Western observers. For Meiji era officials, comparison was a material world-making practice, one entangled with the transformation of the island's conjoined human and more-than-human relations. These processes, and the arrangements they produce, are what I call "landscapes, by comparison" (Swanson 2018). The phrase is inspired by *Anthropology, by Comparison*, an edited collection by Richard Fox and Andre Gingrich (2002), who posit that practices of comparison have been foundational in the making of anthropology as a discipline. Yet here, landscapes, instead of a scholarly field, open up comparison as an ethnographic object in addition to an analytical act.

When anthropologists consider landscapes, cross-cultural comparison is not typically the first topic that comes to mind. Although environmental historians and cultural geographers have examined landscapes as global assemblages, for many, landscapes still conjure a sense of the "local"—of either Indigenous knowledge or traditional rural lifeways—of wisdom that sits in places (Basso 1996). Furthermore, there remains a tendency in popular usage to think of landscapes as more or less self-contained places with ties to particular cosmologies. Speaking of "Japanese" landscapes, for example, often conjures temple gardens and so-called Eastern aesthetics of nature. Paying attention to Hokkaido, however, shows us the utter impossibility of seeing landscapes in such ways. There, we see Japanese landscapes that are made not through some holistic and internal Japanese logic but by comparisons that link the island to geographically far-flung places. We meet Hokkaido landscapes whose species configurations and histories of management cannot be understood separately from specific comparisons with the American West and with particular locales within it.

Somewhat counterintuitively, in the case of salmon, it makes sense to study fish in relation to landscapes, as well as water. Although salmon spend much of their adult lives feeding in the open ocean, they spend the beginnings and ends of their lives in small rivers that are intimately connected to the lands that surround them. During their freshwater phases, salmon are highly sensitive to the variations in water and stream morphology that land use changes generate. Dams can divert water for irrigation and block salmon migration, agricultural runoff can pollute rivers, and logging-related erosion can cause rivers to fill with silt, smothering eggs and degrading habitat. Indeed, nearly any changes to landscapes or rivers can reshape salmon behaviors, modify patterns of fish survival, and rework the genetics of salmon populations. Because landscape processes are literally written into the bodies of the fish, one cannot understand salmon without attention to them. As Hokkaido's agricultural and industrial development seriously damaged salmon spawning habitat, the island's colonization officials also began directly targeting the region's salmon for modernization, remaking both its fishing industry and its fish populations to resemble those of the Columbia River basin, along the US West Coast.

SEARCHING FOR COMPARISONS

When Japanese government officials initially sought to colonize Hokkaido, they were perplexed about what to do with what they perceived as an alien landscape, a place incomparable to Honshu, home to the centers of Japanese political power and cultural identity. The chasm they felt between Honshu and Hokkaido was more than ideological. Biologically and climatologically, Hokkaido is indeed different from the rest of the archipelago. The Tsugaru Straits, which separate Hokkaido's Oshima peninsula from northern Honshu, are so extraordinarily deep (at least 132 meters) that they have largely blocked the exchange of non-avian animals and non-avian-borne plants between the islands (Kondo 1993, 76). During glacial eras, Hokkaido was regularly connected by a land bridge to Siberia via Sakhalin Island, while Honshu, Kyushu, and Shikoku were intermittently linked to the Korean Peninsula. When sea levels were low, mammoths migrated southward from Siberia to Hokkaido, while monkeys moved northward from continental Asia to the other islands. But the Arctic species assemblages that came from Siberia and those from more southerly parts of Asia did not meet and mingle on the Japanese islands. Despite being separated by a mere twenty kilometers, the watery abyss of the Tsugaru Straits kept the non-volant species of Hokkaido and the other islands apart, fostering distinct ecologies to the channel's north and south.[8]

Hokkaido's climate, too, differs from the rest of Japan. Although Hokkaido's major cities sit at approximately the same latitudes as Portland, Oregon, Toronto, Canada, and Rome, Italy, their winter weather is much more extreme than their coordinates suggest. In contrast to central Honshu, where most weather comes from the maritime tropics, Hokkaido's weather sweeps down from the frigid mountains of Siberia and Manchuria. In the winter, these Arctic winds pick up moisture as they cross the Japan Sea, dumping an average of about six meters of snow on Sapporo. When I lived in Hokkaido, I put on my long underwear in late November and did not take it off until mid-April. Along the northern Hokkaido coasts, where salmon fishing flourishes, sea ice drifts across the Sea of Okhotsk and packs against the shore, the ocean groaning as the white ice cracks and shifts. Summer, too, is different in Hokkaido. The 20°C (68°F) summer isotherm, a temperature line that typically marks a boundary between cool temperate regions and warm temperate regions, runs through the Tsugaru Straits (Yabe 1993, 38).

As Meiji era officials formulated development plans for Hokkaido, they looked for models abroad that would offer guidance for such a different place.[9] The Iwakura Mission, a group of Japanese ambassadors and students who took an extended around-the-world study tour in 1871–73, strongly recommended using England as a general model for Japanese development. To the members of the mission, the geography, climate, and culture of the British Isles seemed vaguely similar to Japan, making it an ideal nation to emulate (Willcock 2000, 979). In Honshu, government officials adopted the commission's recommendations, inviting a number of British experts to provide advice on the construction of railroads, telegraph systems, and lighthouses, as well as to establish Komaba Agricultural College, a training school that later became a part of Tokyo University (Russell 2007, 111; Willcock 2000). But in Hokkaido, Kuroda Kiyotaka sought to make a different kind of comparison.

Kuroda, a former samurai from Kyushu, was appointed to the Kaitakushi (also known as the Hokkaido Colonization Commission) in 1870. During the Meiji Restoration, he had distinguished himself by leading imperial military forces against a group of Tokugawa loyalists who had fled to southern Hokkaido in 1869 and briefly established an independent state. By subduing these remaining shogun supporters, Kuroda secured Hokkaido for the Meiji government. Once his military career ended, Kuroda turned to diplomatic and political pursuits, including the settlement of Hokkaido. In 1871, at the Japanese government's request, he traveled to the United States and Europe a few months ahead of the Iwakura Commission.[10] While England might provide a model for mainland Japan, Kuroda saw American agricultural landscapes as a much better template for Hokkaido development (Harrison 1951, 136; Russell 2007, 6). In contrast to the British, the Americans were more experienced in opening new territory, dealing with more severe climates, and cultivating cold-resistant crops. Writing in 1893, Nitobe described Kuroda's decision to compare Hokkaido to the United States:

He saw that the fertile virgin soil could be made to yield its richest treasures only under wise management. But where should he seek wisdom? Japan had long since forgotten the art of breaking up new land; her agricultural system was too intensive to be applied to a newly-opened country; her mining operations were too primitive to be followed on an extensive scale. In General Kuroda's mind there was one source whence he could expect wisdom and knowledge pertaining to new settlements; and that was America. Thither, therefore, he himself proceeded in the fall of 1870. He studied the rapid and wonderful progress of colonization in that country, and thought that the *modus operandi* at work there might well produce similar results in Japan. (Nitobe 1893, 2–3)

Though Nitobe wrote this description in English—likely with rhetorical embellishments targeted toward American audiences—this depiction of Kuroda's fascination with the United States seems more or less accurate.[11] During his visit to the United States, Kuroda was intrigued enough by American settlement practices that he recruited General Horace Capron, the sitting federal commissioner of agriculture, to resign his post and travel to northernmost Japan to serve as an advisor to the Kaitakushi beginning in 1871.

Capron was an established and internationally minded advocate for "modern" and "scientific" agriculture. After the end of the US Civil War, he gained renown for promoting crop diversification in the American South, especially for encouraging farmers to plant citrus trees in addition to cotton (Russell 2007, 81). Even before he headed to Japan, Capron was thinking and acting beyond the boundaries of the United States. He had become a corresponding member of the Society for the Promotion of National Industry of Brazil to become more familiar with South American crops. Learning of the success of seedless oranges through the society's materials, he arranged to have two of the trees shipped to California, an act that sparked the West Coast navel orange industry. Additionally, in 1869, he started an international seed exchange program, inaugurating it by shipping 130 seed packages to the new Meiji state (Russell 2007, 81).

When he arrived in Japan in 1871, Capron stayed in Tokyo for many months, crafting his recommendations and plans for Hokkaido before he ever set foot there. Capron also established experimental government farms in Tokyo to provide a way station for plants and animals in transit from the United States to Hokkaido (Fujita 1994, 36). The Honshu farms served not only as sites for research and acclimatization but also as places to publicly display the Kaitakushi's progress to Tokyo-based leaders. In 1873, the emperor himself came to inspect the farm's crops and animals (Walker 2004, 257). In mid-1872, Capron moved northward to Hokkaido itself, and upon arrival, Capron was impressed with its potential:

> This island is just wonderful. Its true value has not been recognized nor regarded as important. Its mineral resources are abundant. Its fishery resources are inexhaustible. Its woods are superior in quality and abundance and its agricultural productive power is great. (Quoted in Fujita 1994, 38)

But at the same time, Capron was disappointed with the island's existing experimental farm, started by a German farmer, which was yielding little produce (Russell 2007, 140). He also found the quality of the island's farm animals to be so dismal that he suggested that the Kaitakushi order "all native stallions, bulls, and boars be either altered, i.e., deprived of the power of generation, or removed to some remote part of the island, and by the introduction of foreign animals in their stead for breeding purposes" (Capron, cited in Russell 2007, 141).[12]

Over the next two years, Capron would spark a revolution in Hokkaido agriculture and land use by introducing American crops and livestock. The lists of species that made their way across the Pacific by steamship is truly impressive. Some came in the form of cuttings—cherries, nectarines, plums, peaches, apricots, raspberries, currants, black gooseberries, strawberries, rhubarb, quinces, and grapes. Others arrived as seeds—onions, turnips, carrots, cabbage, lettuce, tomatoes, beets, celery, spinach, corn, peas, beans, and potatoes. Still others arrived on the hoof—Devon and Durham cattle, Berkshire and Suffolk pigs, Cotswold, Merino, and Southdown sheep, and Arabian horses. Their numbers were not small; by the end of 1873, 32,775 young fruit trees had been shipped to Hokkaido.[13] In total, 224 varieties of fruits and vegetables made their way to Japan under Capron's tutelage (Russell 2007, 129, 132, 134).

Capron also recruited additional Americans to assist his efforts to help the Kaitakushi transform Hokkaido's landscapes. The cadre of American men that the Japanese government hired at his recommendation surveyed the island, mapped its geology and rivers, laid out the grid system for its capital city, built mechanized sawmills, fostered the development of mining industries, and helped with road, bridge, and railroad construction (Duke 2009; Fujita 1994). One of these foreign pioneers was Edwin Dun, an Ohio rancher, whom Capron selected to introduce modern livestock production to northernmost Japan. Beef eating, in particular, was framed as an act redolent of modernity. The Japanese government began to heavily promote meat consumption, arguing that Europeans had strong, muscular bodies because they regularly dined on mammal flesh.[14] In 1872, the royal household announced that the emperor regularly ate beef and mutton (24). According to Japanese food studies scholar Katarzyna Cwiertka, in the nineteenth-century West, meat eating was perceived as a source of national strength and linked to social Darwinism: "A leading British scientific publicist . . . stated in one of his lectures of 1860 that 'those races who have partaken of animal food are the most vigorous, most moral, and most intellectual races of mankind.' Similarly, an American cookery writer . . . argued that the British dominance of India proved the fact that meat-eaters dominated world politics." In a moment when such sentiments circulated alongside new Western notions of nourishment and sanitation, the Japanese government quickly added canned beef to their military menus (Cwiertka 2006, 33, 63–64).[15]

Hokkaido's Kaitakushi was interested in the economic value of such animals. For the island's colder and more marginal climates, livestock rearing seemed more promising than rice farming. Dun, with years of practical experience in the US Midwest, became their guide. He brought more than one hundred cattle and one hundred sheep to Japan, including some from his own farm (Hokkaido Prefectural Government 1968, 44–45). But once he arrived in Hokkaido, he faced a serious challenge: the island was no pastoral paradise. Its grasses were poor, its farms lacked fences, and wolves prowled its mountains. Dun and the Kaitakushi set out to make the landscape safe and hospitable for the animals that symbolized modernity. They introduced Kentucky bluegrass, red top, timothy, and clover; they built miles of split-rail fences; and they exterminated wolves and wild dogs with strychnine, a chemical poison widely used for predator control in the western United States (Fujita 1994, 60; Walker 2004). The practices worked; they helped to build beef, dairy, and horse industries in Hokkaido, while decimating the island's canid populations. They successfully turned miles of hills and plains into parcels of pasture.

As in the case of the American West, the Kaitakushi and their American advisors sought to exterminate not only the animals but also the Ainu culture that impeded their agricultural plans. Capron, who had served as a federal Indian agent earlier in his career, was an advocate of so-called native assimilation policies, which sought to eliminate Indian lifeways through forced agriculture (in contrast to other officials who favored policies that created small reservations and restricted American Indians to them). Although he expressed some remorse about the brutal treatment of Indians, Capron participated in the Indian resettlement and likely promoted Indian farming (Medak-Saltzman 2008, 100–102). Although Capron's role in Ainu policy is unclear, the Kaitakushi (and later the Hokkaido prefecture government) adopted strategies that share some similarities to the assimilation-focused US Indian policies that were in vogue after the US Civil War, such as boarding schools, forced agriculture, and dubious land allotment schemes.[16] For example, in 1872, the Kaitakushi pressured thirty-seven members of the Ainu elite to attend a temporary school in Tokyo, where they were taught agriculture and livestock farming with the hope that they would take such skills back to their villages and inspire other Ainu to adopt farming lifeways (Frey 2007, 69–96).[17] From 1901 to 1937, the government operated a segregated system of Ainu schools, where children, who were forbidden to speak their native languages, took coursework in Japanese, arithmetic, farming (for boys), and sewing (for girls).[18]

Japanese Ainu policies also seem to have been influenced by the legal maneuverings that the US government used to disenfranchise American Indians. Declaring Hokkaido empty land and instituting a new property-rights regime, they stripped Ainu people of their lands. In 1899, as part of the Former Aborigines Protection Act (Kyūdojin Hogohō), they created land allotment practices that echo parts of the 1887 Dawes Act, which turned Indian lands into privately owned farmsteads (Medak-Saltzman 2008, 103–5). The Kaitakushi further forced Ainu people into exclusively agricultural ways of life by strictly enforcing hunting and fishing bans that deprived the Ainu of access to critical food supplies. In 1876, the Japanese government outlawed the bows and poison-tipped arrows that Ainu people used to hunt deer. Three years later, the government prohibited the freshwater capture of salmon and trout (Aoyama 2012, 119). The aim of such laws—as for many contemporary US policies—was the functional elimination of Indigenous lifeways. As Ainu leader Kayano Shigeru (1926–2006) described it, the "law banning salmon fishing was as good as telling the Ainu, who had always lived on salmon, to die. For our people, this was an evil law akin to striking to death a parent bird carrying food to its unfledged babies" (Kayano 1994, 58–59).

Although the Kaitakushi did not bow to American advice and friction with the foreigners was not uncommon, the advisors undoubtedly spurred shifts in the Kaitakushi's approaches to Hokkaido's social and natural landscapes. However, they did not stay long, as the Kaitakushi hired most of them on one- to three-year contracts. Furthermore, in 1882, when the central

Japanese government reorganized Hokkaido's administration—replacing the Kaitakushi with another form of central governmental control—most of the directly employed foreigners were sent home (Hokkaido Prefectural Government 1968, 26).[19] But another institution, the Sapporo Agricultural College, continued to work outward from American-inflected logics of modern scientific agriculture and natural resource management, expanding them to transform Hokkaido's lands and waters for decades to come. Immediately after joining the Kaitakushi, Capron began advocating for the development of an agricultural school in Japan. Kuroda and others were easily persuaded. As Nitobe would later write, "The simple adoption of American methods without trained hands to rightly direct them, would merely amount to an apish trick" (1893, 3). Japan needed people who could both inhabit modernity's subjectivities and perfect its technical practices. Education was thus a key facet of the Kaitakushi's efforts.

In 1875, Kuroda asked the Japanese ambassador in Washington, DC, to secure the services of an American educator capable of establishing a first-rate agricultural college in Hokkaido. Several years earlier, the Kaitakushi had attempted to build a temporary school in Tokyo for the education of modern farmers, but the institution had been disorganized, and it was deemed a failure (Duke 2009, 201). Kuroda wanted American advisors who could turn their floundering school into a full-fledged institute of higher education. The Japanese government managed to recruit a consultant of the highest caliber—William Smith Clark, then president of the Massachusetts Agricultural College (MAC). MAC was one of the first land-grant colleges founded under the Morrill Act of 1862, which provided funding for schools where "the leading object shall be, without excluding other scientific and classical studies, and including military tactics, to teach such branches of learning as are related to agriculture and the mechanic arts" (US Congress 1862). Clark, one of MAC's founding members, embraced this challenge and sought to create the United States' first generation of well-trained specialists in scientific agriculture. When he was invited to create a similar college in Japan, Clark jumped at the opportunity, taking a year's leave from MAC to travel to Hokkaido. In a letter to his wife, Clark remarked on this exciting opportunity to "rebuild M.A.C. with variation and possibly some improvements on the other side of the earth" (cited in Willcock 2000, 987).

In summer 1876, Clark arrived in Hokkaido along with two other MAC professors, William Wheeler (civil engineering and mathematics) and David Penhallow (chemistry, botany, agriculture) (Fujita 1994). Immediately, they began creating Sapporo Agricultural College (SAC). One of Clark's first requests was that the Kaitakushi build a model farm, then turn its ownership over to SAC for use in agricultural education (Kataoka 2009, 6-1). Per the Americans' suggestions, the new facility included both crop production areas and a dairy barn, which also included spaces for horses and pigs (6-3). Originally, the new farm building had a descriptive name: the Delivery Room and the Stable; Clark, however, renamed it the Model Barn to symbolize its intended role as a template for modern agriculture in Japan (6-3). The curriculum that Clark created for SAC embodied the spirit of the Morrill Act, emphasizing practical education and military training, but not at the expense of more scholarly pursuits.[20] In the school's early years, the courses included geometry, English, German, elocution, and political economy, along with drainage and irrigation, manures and crop rotation, vegetable pathology, stock farming, and veterinary science. Notably, students also took classes titled "History of Colonization" and "Political History of Europe" (see Nitobe 1893, 35–42, for a complete list of courses). Natural history, and its mode of scientific nature observation, was also a critical part of the curriculum. Faculty took students on scientific expeditions around Hokkaido to collect specimens and to teach the young Japanese to see the world through the lens of natural resources management. During its second year of operation, the school added a natural history museum so that its students could more easily make comparisons by viewing "the natural history of Japan and its productive resources, together with such specimens as may be obtained from abroad by purchase or exchange" (Sapporo Agricultural College 1878, 2; Yaguchi 2002, 104).

At SAC, Clark collaborated with Japanese students to create a school with much more ambitious goals than simple instruction in the cultivation of crops, the proper siting of mines, and the preservation of botanical specimens. Through the study of agricultural practices and natural resource management, he sought to cultivate modern male subjectivities. The goal of the school was to create an improved breed of men, alongside better breeds of wheat and horses; the school wanted to make men who could become leaders of a societal shape-shifting, an agricultural and industrial revolution in the service of modern nation-building. According to the school's second annual report, in an introductory letter penned by then college president William Wheeler to Kuroda Kiyotaka, the head of the Kaitakushi, Japan would need its own legions of "agricultural and industrial exhorters" to "induce" common farmers to accept the "privileges" of modern agriculture, to make them "understand, or to have faith, that their present condition and that of the country could be made better through such radical innovations" (Sapporo Agricultural College 1878, 19). As Wheeler

Sapporo Agricultural College Model Dairy Barn, located on the campus of what is now Hokkaido University. Completed in 1877, the barn was inspired by a similar structure at Massachusetts Agricultural College. Photo by author.

continued, "To furnish men for missions of this nature should be considered one of the first objectives of the Agricultural College" (20).

Clark and later staff made self-cultivation and moral education core educational goals. To spread the gospel of modernity, SAC students were to first inhabit its subjectivities themselves. They were to develop what the school called "frontier spirit." Although SAC's moral education was not reducible to religion, it certainly included sizable doses of it. When Clark ran the school, every morning before lecture he led the students in a hymn, a scripture reading, and a recitation of the Lord's Prayer. With Clark's encouragement, the entire first-year class signed his "Covent of the Believers in Jesus," converting to Christianity, and other SAC students reportedly practiced worship forms in their limited spare time: "The boys in [a second-year student's] group took turns as a 'pastor' and rotated the meetings among their college dorm rooms. Whoever was the minister for the week brought in an empty flour barrel to serve as a pulpit, which was draped in a blanket. Blankets were laid on the floor for the 'congregation' while the appointed minister sat in the sole stool" (Czerwien 2011, 29, 36).

SAC Christianity was eclectic and predominately lay-led. One student wrote that "it was interesting [to us] because it was a practical religion, unlike that taught by ordinary missionaries. It was religion without the odor of religion" (Maki [1996] 2002, 178). Yet at the same time, SAC Christianity was deeply Protestant in its ability to link self-cultivation to national development and capitalist-oriented progress. Protestant Christianity provided important frameworks for comparison-making in part due to its focus on "improvement." For the Americans in Sapporo, modernizing one's soul was inseparable from—and critical to—improving one's country.[21] Understanding what "improvement" might be and whether or not one had accomplished it was seen as an inherently comparative task, with American Protestant Christian teaching offering up frameworks of heathen/Christian and backward/modern alongside ideal models, ranging from Jesus to mechanized farmsteads. The SAC instructors, New Englanders steeped in liberal education, believed that the students needed to be inculcated with desire, with yearnings for continual improvement at the scale of both the self and nation. When Clark was departing Sapporo at the end of his tenure at SAC, he reportedly shouted his most important advice to his students as he trotted away on horse: "Boys, be ambitious!" More than a century later, the phrase continues to be well known throughout Japan and was prominently featured in a mobile-phone advertisement that played incessantly on television in the late 2000s, when I was studying and researching in Japan. SAC instructors, including Clark, felt that such subject formation required far more than "book learning." Students were required to take a course called "Manual Labor," to perform gymnastics, and to regularly engage in hands-on activities. SAC also used school meals to craft students who would be at home with one foot in the East and one in the West. In addition to Japanese-style rice-based meals, the students were introduced to Western-style staples, such as chicken, venison, coffee, bread, butter, and ice cream, served on flat plates.[22]

In total, such practices appear to have had their intended effects. By 1898, when a labor activist wrote the following words, SAC and its graduates had already begun to draw attention in Japan:

> It is the only college in Japan that has the so-called 'college spirit' which has been moulding the character of students ever since the distinctive impression made upon the college by the first Pres. W.S. Clark. The college is noted for making men though she has not neglected making scholars. Sons of the

college are conspicuous figures everywhere throughout the
Empire. (Sen, quoted in Willcock 2000, 991)

Although the school was located on the margin of Japan, it was one of
the fledgling nation's most important nineteenth-century institutions
of higher education, and its early graduates joined Japan's first generation
of cosmopolitan gentlemen.[23] As a result of their Western educations, the
early cohorts of SAC students became unusually skilled in various comparative practices and were strongly represented within an emerging group
of Japanese cosmopolitans. They became Japan's translators, negotiating
across languages and concepts. They went on to earn advanced degrees
from top institutions in the United States and Europe, including Harvard,
Cornell, and Johns Hopkins. They became diplomats and statesmen. One
rose to the position of prime minster, another to that of under secretary-
general of the League of Nations. While citing knowledge they gained in
Sapporo, they guided Japan's colonization of Taiwan and Korea, suggesting
plans for their agricultural development. One became the chancellor of
Tokyo University, while many others also took up teaching, fulfilling SAC's
dream that they would spread new knowledge and a new spirit across
Japan (for all these examples, see Willcock 2000, 1016). About 40 percent
of the students who graduated between 1880 and 1895 became teachers "for
a substantial part of their working lives" (1016). Some become prominent
Christians, starting a church in Sapporo and a small religious movement in
mainland Japan. They introduced Nathaniel Hawthorne to Japan, developed a Shakespearean theater, authored bilingual dictionaries, established
a fine arts school, founded English language newspapers, and published a
Japanese magazine called *English Youth* (1015).

Yet as they became citizens of the world, the school's graduates did not
neglect Hokkaido, enacting their new transnational philosophies on the
island's landscapes. During the school's early days, students were required
to sign a pledge committing themselves to serve the Kaitakushi in its efforts
to develop the island: "After graduation I will become a citizen of Hokkaido
and will serve in the Colonial Department for five years upon the same terms
as other officers of similar rank" (Dudden 2005, 10–11; Sapporo Agricultural
College 1878, 94). Although the public-service requirement was soon
dropped, more than a third of the school's pre-1900 alumni remained in
Hokkaido permanently, becoming the leaders of its businesses and institutions. "The Society of the Advancement of Agriculture, the Fishery Association, the Natural Science Society, a body called the Friends of Learning,

the Pomological Society, the Economic Club, the Young Men's Christian Association, the Temperance Society, the Silk Culture Association, and many other minor organizations all count among their most active members and promoters the graduates of the College" (Nitobe 1893, 30). Although the American advisors are often given much of the credit for the island's colonization, the SAC graduates were the ones who more substantially transformed Hokkaido. The logics and practices they preached and performed set off a cascade of landscape changes. They drained the marshlands, converting them to agricultural land. They cut forests and processed wood products. They built coal and gold mines. With their influence, from 1869 to 1912, Hokkaido's population increased from about fifty-eight thousand to 1.7 million (Ivings and Qiu 2019, 291).[24] Throughout the twentieth century, the island's development continued on a relatively consistent path of agricultural intensification, natural resource exploitation, and industrialization, despite the vast changes and turbulence of war and, later, of postwar growth—with Hokkaido's landscapes straining under the changes.

COMPARING SALMON

While agriculture was an important focal point, the colonization of Hokkaido was more than a terrestrial process. It marked the transformation of rivers, oceans, and fish as much as it did the transformation of lands. Although many of the Kaitakushi's projects focused on establishing scientific agriculture in Hokkaido, colonial administrators did not overlook the direct modernization of its seas. From the beginning, fisheries were considered to be one of Hokkaido's most valuable assets. Hokkaido's fecund fishing grounds were what initially drew ethnic Japanese interest in the island, and in 1891, after more than two decades of state agrarian encouragement, more than 70 percent of the island's population still worked in fishing-related employment (Irish 2009, 132).[25] While the American advisors who arrived in Hokkaido tended to initiate a more land-oriented colonization, Japanese officials, who were not about to neglect the fisheries that had long been the region's economic mainstay, put substantial effort into their development.

For hundreds of years prior to the Meiji Restoration, ethnic Japanese people consumed sizable quantities of Hokkaido's salmon and herring, first by trading with the region's Indigenous Ainu people and later by forcing Ainu to labor for Japanese fishing firms. During the Meiji period, however, colonial administrators began to see Hokkaido's seafood as a potentially lucrative export in addition to a domestic foodstuff. By the mid-1870s,

Hokkaido bureaucrats expressed a strong interest in establishing a canned salmon industry. Only a decade earlier, in 1864, two fishermen from Maine, the Hume brothers, became the first people to try to can Pacific salmon. The men had moved to California as 49ers, but when they did not strike gold, they turned to silvery fish. They established an experimental cannery along the banks of the Sacramento River and began packing salmon into handmade metal tins. Their first products were such a success that they decided to relocate to a location with larger salmon runs and better possibilities for expansion: the lower Columbia River, along the border of Oregon and Washington State. In 1866, the Hume brothers built a small cannery at Eagle Cliff, Washington, near the mouth of the Columbia. In their first year, they sold only four thousand cases, but in their second year, their sales more than quadrupled, to eighteen thousand cases. This success was a marked change from pre-canning attempts at commercialization. From the 1830s to the 1850s, the river's immense salmon runs captured the attention of white explorers and businessmen, who tried packing the fish in salt and brine. Such methods, however, failed to turn a profit, because too much of the salmon spoiled en route to major markets along the US East Coast. Canning technology, however, created new trade routes by suspending time (Naylor 2000). With salmon safely preserved in metal vessels, Columbia River fish could be shipped to markets anywhere in the world. Customs records from 1873 show that Columbia River salmon were already being directly exported to England, China, and Australia, and by 1875, Astoria, a port city at the river's mouth, had become the center of a global canned seafood industry with twenty-four foreign and domestic ships taking on cargoes of canned salmon. In this newly transnational form, the industry rapidly grew; in 1873, there were eight canneries dotting the banks of the lower Columbia, yet only ten years later, the number had increased to thirty-nine (Penner 2005, 10; Tetlow and Barbey 1990, 5, 6, 8).

The Columbia River salmon industry created a buzz among entrepreneurs on multiple continents.[26] When Hokkaido administrators heard about it, they thought that they might be able to establish something similar in northern Japan. Snippets of correspondence from 1876 and 1877 indicate that the Kaitakushi were beginning to think about the potential export value of their salmon. Japanese government officials yearned for a favorable balance of trade in order to rapidly acquire foreign currency. In addition to developing an export-oriented silk industry, Japanese government officials began to consider the possibilities embodied in fish. In late 1876, the Kaitakushi began sending samples of salmon—both smoked and experimentally canned—to foreign merchants and diplomats for evaluation. One Yokohama-based merchant named J. D. Carroll was optimistic enough about the test products he received that he sent a reply to the Kaitakushi in January 1877, reporting that he found their tinned salmon to be "fine" and their smoked salmon to be "excellent" (Carroll 1877). He thought the products might do well if exported to China and offered to do business with the Kaitakushi in the future. The Kaitakushi, however, had more high-prestige markets in mind. They sent several samples of smoked salmon to US consular staff along with a letter asking the Americans to report back on "how it suits your American taste" (Yasuda 1876). In addition, they wrote a memo to William Clark asking what part of the United States he thought might provide the most promising market for Hokkaido salmon (Kuroda 1877a). But many Americans and Europeans were less than enthusiastic about how Hokkaido salmon would fare in their stores. The Tokyo-based representative of a London-based trading firm reported mixed reviews of the first batch of Hokkaido salmon. The British thought that the smoked salmon was decent, but "continental" tasters found it "mouldy and greasy" (Ahrens 1877b). No one liked the canned fish: "As to the sample of tinned salmon sent, the reports both from London and the Continent are unsatisfactory. The salmon on arrival were found broken into small pieces and the color had turned bad and it could not be brought into competition with the preserved salmon from America" (Ahrens 1877b, 2).

Undaunted, the Kaitakushi moved forward with their plans to commercially produce and export canned salmon. In June 1877, Kuroda asked Capron to "employ one practical man well acquainted with the precepts of making canned salmon etc for term of six months" (Kuroda 1877b). Capron secured the services of Ulysses S. Treat, a cannery man from Maine, who arrived in Hokkaido later that same year along with an assistant named T. S. Sweat (Cwiertka 2006, 62). When he saw Hokkaido's fall run of fish, Treat was enthusiastic about their commercial potential, boasting that "the salmon fishery in the Ishikari River is one of the largest yet known. It is stated that, in some seasons the catch amount is about 1,800,000 fish" (Treat 1878). At the Kaitakushi's request, Treat oversaw the construction of a cannery near the mouth of the Ishikari River and provided instruction in canning techniques. Under his direction, the cannery produced 12,092 two-pound cans of salmon in its first year, in addition to a few cans of salmon eggs, some barrels of pickled salmon, and a bit of smoked fish (Treat 1878).[27]

Aiming to impress Europeans and Americans, the Ishikari factory wrapped these first cans in bright red bilingual labels, similar to those in use on the Columbia River, with directions for use in both English and Japanese (see photos below). The kanji characters on the label may have added an alluring Oriental mystique for overseas audiences, but they served little practical function.

Although Japanese people ate sizable amounts of fresh and dried fish, Hokkaido canned salmon were never intended for domestic markets. They were too expensive for Japanese consumers and rather unappealing to Japanese palates.[28] Instead, the Kaitakushi consistently courted European tastes, seeking feedback on their evolving product from white foreigners. A Kaitakushi official sent some of the 1877 salmon to the Japanese Consulate office in Marseille, France, with a request "to distribute the salmon to some Europeans, who are doing the business with, and give me their opinions as well as your own of its quality and also furnish me the information of its sale in Europe, for we have the intention to promote this enterprise to a great extent" (Yasuda 1878). Yet this new batch of Hokkaido salmon still failed to match the flavors and textures for which European taste buds yearned. One French trader could find nothing he liked about Hokkaido fish. The salmon "was not a first class fresh" fish, the "boiling was too long," and the season in which the fish was prepared was likely "not proper" (Freres 1878). In his opinion, even the size and shape of the tins was wrong. The Dutch ambassador to Japan also discouraged the Kaitakushi from trying to sell their fish in Europe, advising that the fish would be likely to find "a better and more profitable market in British India and Java" (Bauduin 1879). A British merchant was impressed by Japanese canning technique but was disappointed by the flavor, texture, and color of the Kaitakushi's product:

> The fish prepared with Japanese salt has a peculiar flavor, which is probably due to that kind of salt. We doubt if this flavor would be liked in Europe. . . . The fish in all the cans, although perfectly preserved was of a very light colour, and in our judgment too dry and tough in texture to be ranked as equal to the Oregon Salmon. . . . We think that the people of Europe, who have become accustomed to the appearance and taste of the Oregon Salmon, would not consider the Hokkaido fish as equal to it either in quality or value. The Hokkaido Salmon is no doubt very good food, but the Oregon fish would probably be much preferred, and it might be difficult, at least in the beginning, to introduce, or to obtain a fair price for, the Japanese product. . . . We would suggest that you should yourself make a comparison between the Oregon and the Hokkaido fish, remembering that the toughness or firmness (hardness) of fibre which in Japan is considered a merit in fish, is not so considered

Columbia River salmon canning label, 1881. Courtesy of the Oregon State Archives.

> in foreign countries, though of course the tenderness of fibre
> which is preferred there must not degenerate into softness or
> rottenness. . . . We regret not being able to give you a more
> encouraging report on your samples, the packing of which
> seems quite faultless.[29] (Walsh 1877)

Like this British merchant, most of the foreigners who sampled Hokkaido salmon compared it immediately with the Columbia River fish that were already gaining international popularity. They urged the Kaitakushi to learn to make the same comparison. A representative from a London trading firm described how salmon were canned in Oregon and recommended that the Japanese obtain "practical experience" in how salmon canning was performed there (Ahrens 1877a).

William Clark also felt that the Hokkaido salmon industry needed to learn from the Columbia River. He wrote letters to the Kaitakushi about the successes of Oregon canneries: "The total amount taken at Astoria and vicinity is estimated at 40,000,000 pounds annually. 25,000,000 cans weighing one pound and a quarter each were sold for about $3,000,000 in 1876. 7,920,000 cans were sent to England. The demand for the salmon is so active that it is all sold before the fish are caught" (Clark 1877a). Clark convinced the Hokkaido officials to pay him 250 gold yen to travel to Oregon and prepare a report for the Kaitakushi on its salmon industry. In summer 1877, when he returned to the United States, Clark made a beeline to Astoria, Oregon, where he drafted a thirteen-page report on the Columbia River salmon

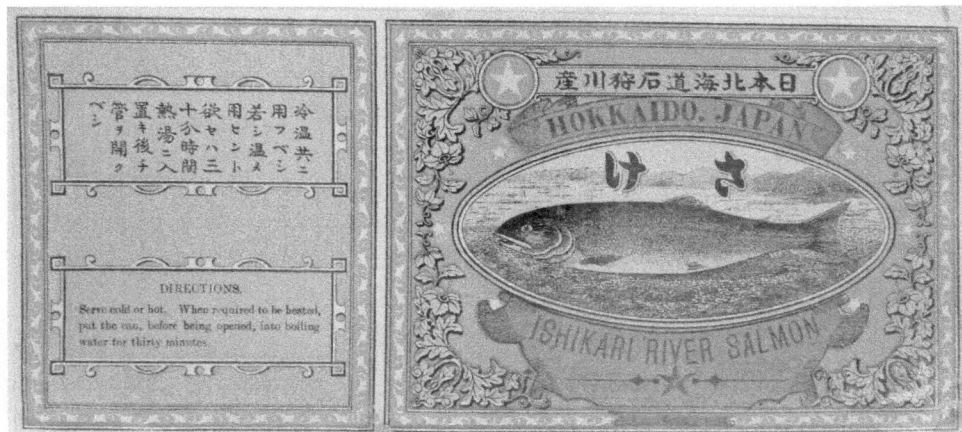

Hokkaido salmon canning label, 1877. Courtesy of the Archives of Hokkaido.

harvest and the region's canneries. He provided a comprehensive overview of an array of topics: gillnet fishing methods, tin can production, practices for killing and bleeding fish, temperatures and diameter of boilers, how to check for defective cans, and how to pack salmon in wood crates. He also wrote about the organization of labor, including the productivity of cannery shift workers and the system through which canneries leased boats and nets to fishermen who lacked the capital to buy them (Clark 1877b). Clark thought that Hokkaido canneries, like those of the Columbia River, would have to seek out British markets: "England takes nearly one third of the [Columbia River] canned salmon and Australia a considerable quantity. Japanese salmon would probably have to seek a market in England or some of her colonies. Only English laborers will buy such expensive food. There can be little doubt however that a good article can be sold at a remunerative price in some part of the wide world."[30]

Soon, Hokkaido's canned salmon did indeed successfully compete with salmon from the US West Coast. By 1910, Japan's salmon industry had taken off, fueled by fish from both Hokkaido and the new northern territories acquired during the 1904–5 Russo-Japanese War. Canning companies quickly expanded into Sakhalin, the Kurils, and even mainland Kamchatka, where at the time, Japan had treaty rights to establish salmon fishing colonies.[31] By 1932, there were ten canneries on Hokkaido and thirty-three more in the Kurils, Sakhalin, and Kamchatka (CFAJ 1934, 25–27). Approximately 80 percent of Japanese-produced canned salmon was exported, and of this exported fish, about 80 percent was bound for England, with the remainder headed to France, Holland, Belgium, and South Africa (31–32). Throughout the early twentieth century, the Canned Foods Association of Japan actively marketed canned salmon products, sending its managing director on an extensive promotional tour of Africa, Europe, and the Balkans in 1930 (102). In 1934, the organization was pleased to report steady increases in exports, "indicative of the fact that Japanese canned salmon has maintained its good reputation in foreign lands" (33). The fish had undoubtedly become "the backbone of the canning industry in Japan" (4).

HATCHERY HISTORY

Modernizing fish, however, meant more than putting them into cans. It also entailed efforts to rationalize nature and increase its productivity. While at the helm of the Sapporo Agricultural College, William Clark suggested that Hokkaido improve its salmon species in the same manner as its horses and cattle: by replacing the weak Japanese stocks with bigger Western versions. He called for

> the introduction into the Ishikari River of the Salmo Salar or large salmon of Europe and America. This species not only grows to a much larger size that the salmon now frequenting Hokkaido, but its flesh is much firmer and better adapted to canning. There would seem to be no special difficulty in bringing the impregnated eggs from the Sacramento River in California and hatching them in the waters of the Ishikari, from which this most valuable fish could then be distributed to all parts of the Empire where the conditions are suitable for its growth. (Clark 1877a, 2)

Although they did not follow such advice to its letter, the Kaitakushi indeed took suggestions about fish culture seriously. In 1877, the same year that Hokkaido officials instructed Ulysses Treat to establish Japan's first salmon cannery, they also authorized him to conduct the island's first salmon hatching experiments. Treat had boasted that fish hatcheries were an integral part of the cannery complex; if hatcheries were properly established, he said, there would "be no doubt of success" for the entire industry. Hatcheries were both the modern way and the American way:

> Millions of salmon eggs are thus hatched in America, every year, and the benefits derived from the operation are already

making themselves manifest, not only in the increasing numbers of fish to be found in places where salmon were formerly abundant and from which they have been driven by excessive fishing, but in their appearance in places where they have previously been wholly unknown. (Treat 1878)

By the late 1870s, Treat was likely preaching to the converted. Japan had a long history of fish culture efforts, and in the Meiji moment, the country hardly needed to be convinced of its potential benefits. Since the 1750s, samurai had been building spawning channels and altering Honshu rivers to boost salmon reproduction (Kobayashi 1980, 96), and by the Meiji era, members of the Japanese government were already enthusiastic about more active and interventionist approaches to fish cultivation. In 1873, several exhibitions at the Vienna World Exposition had caught the eye of a Japanese official in attendance. One was the Australian delegation's hatchery exhibit. It explained how beginning in 1864, salmon and trout eggs had been successfully shipped from England to Tasmania, where they had been hatched and released into Australian rivers that had never before borne salmon. This report of successful of salmon propagation captivated the Japanese official, but from the information provided in the exhibit, he could not quite understand the exact techniques.[32]

While fish culture was not itself novel to Japan, large-scale salmon production practices emerged from a variety of new assemblages. The Sapporo Agricultural College played a primary role in the building of the island's modern fisheries, just as it did for its land-based agriculture. From 1878 to 1887, John Cutter, a Massachusetts doctor, taught a variety of courses at SAC, including zoology, veterinary medicine, and fisheries sciences (Minamoto 1993, 27). Although there was no fisheries department during SAC's first decade, the school still inspired some of its earliest students to think about the scientific management of the seas. Uchimura Kanzō, a member of the second graduating class, essentially majored in fisheries and gave a graduation speech titled "Fisheries Is One of the Sciences" (Matsuda 2002, 407). The fisheries curriculum grew quickly; courses in ichthyology and fishing gear and methods were added in 1884, a class in aquaculture in 1887, and another in fisheries science in 1889 (407). In 1906, the school formalized its commitment to training managers of the sea by creating a separate Department of Fisheries. Ultimately, the college dominated fisheries education in Japan for more than a century. Until 1987, Hokkaido University (SAC's successor) offered the only fisheries science doctoral program in Japan (408).

As was the case with agricultural development, SAC graduates pioneered fish cultivation practices in Hokkaido. Ito Kazutaka, a member of SAC's first graduating class, revolutionized Hokkaido's fisheries by instituting the salmon hatchery system that remains the backbone of today's salmon industry.[33] Throughout his life, Ito hewed to a path typical of SAC graduates: he converted to Christianity, helped found a church, and became vice president of the Japan Temperance Union. But in contrast to many of the other graduates, he sought to ranch Hokkaido's seas rather than till its soils. After his graduation from SAC, Ito accepted a post with the Kaitakushi to fulfill the school's government-service requirement, and Ito, like many of his classmates, turned this mandatory service into a permanent career as a public official. When the Kaitakushi was converted into a prefectural government, Ito became the head of Hokkaido's first prefectural fisheries department (*suisan kachō*). In 1886, at the request of the Japanese government, he traveled to North America to study US and Canadian fisheries practices, with the aim of improving those of Japan's north. During a twelve-month whirlwind tour, Ito traversed the continent, visiting more than fifteen states and provinces. He met with US officials in Washington, DC, toured New York City's Fulton Fish Market, visited fish processing plants in Rhode Island, and made careful observations of New England's cod fishery.[34]

Ito's most important activities were centered on salmon. He traveled to Bucksport, Maine, to document the practices of a brand-new institution: the salmon hatchery. Fish culture there, like everywhere in the United States, was still in its infancy. Maine's inaugural salmon hatchery was not constructed until 1871, with the Bucksport facility following a year later. This was one of the few places where Ito could observe such novel practices of producing fish. When Ito was touring the continent in 1886, it would have been impossible for him to visit a Columbia River hatchery, for a simple reason: salmon hatcheries had yet to take root in the Pacific salmon heartland. Although the US Fish Commission had established one small hatchery on Oregon's Clackamas River in 1877, the facility had closed in 1881 from lack of funding and was not reopened until 1888 (Northwest Power and Conservation Council n.d.).

Yet while the East Coast was advanced in terms of hatcheries, the West Coast was the world leader in canneries. So after his visit to Maine, Ito traveled by train first to British Columbia's Fraser River, then to the mouth of the Columbia. He timed his trip perfectly, arriving on the West Coast in mid-September, when the region's rivers swarmed with salmon. As a guest of an Oregon fishery official, Ito spent a week observing various parts of the mainstem Columbia River. In the river's middle reaches, he watched American Indians harvest salmon, while near its mouth, he surveyed commercial fishing techniques and toured a cannery. By the time Ito returned to Hokkaido, his notebooks were filled with meticulous and detailed line drawings of hatchery incubators, his mind racing with new ideas. Modern fisheries science was still so embryonic in North American that it stood in sharp contrast to agricultural pursuits, where Hokkaido tended to appear "behind" the West. Ito's job was not to help Hokkaido "catch up"; it was to help the island join in—and perhaps even lead—the mounting wave of late nineteenth-century fish culture. In 1888, the same year that the first Columbia River hatchery reopened, Ito established Hokkaido's Chitose Central Salmon Hatchery, modeled after Maine's Bucksport facility (Kaeriyama 1989, 627). As Ito continued to experiment with fish cultivation and expand Hokkaido's hatchery system, he was, if anything, ahead of the curve. With his inspiration, Hokkaido's fish cultivation program grew to a network of fifty hatcheries in twenty years, a pace faster than that found along the US West Coast (Kobayashi 1980, 97).

Ito was clearly not an imitator but an innovator. For example, he combined the design of a fish wheel that he saw on the Columbia River with Japanese weir technology to create a new method for harvesting fish hatchery brood stock.[35] Ito seems to have strongly felt that such modernization and innovation required comparative thinking. He founded the Hokusui Kyōkai, a fisheries society that shared information about evolving technologies. On his return, the group published Ito's report from his North American fisheries study tour, making it widely available to those in the industry. As a result of his technical innovations and dissemination efforts, Ito was—and still is—hailed as the father of modern fisheries in Hokkaido.

Yet the development of modern fisheries science and managerial control in Hokkaido, as in other sites around the world, was far from a story of unambiguous success. Hatcheries were quite intrusive; workers would typically build weirs that spanned rivers bank to bank, funneling all migrating salmon into holding pens for hatchery use. The method, which blocked most salmon from swimming upstream and spawning on their own, essentially converted a given river's salmon from natural spawning to an allegedly superior mode of reproduction. Salmon hatcheries allowed Hokkaido's fisheries managers to feel modern, but they did little to boost salmon populations. Although hatcheries released large numbers of juvenile fish, most of the hatchery smolts are thought to have died soon after they reached the ocean, as salmon harvest numbers showed no increases as a result of early hatchery efforts. Ito's Chitose hatchery and other Hokkaido facilities diligently researched salmon biology, but they still had much to learn. Not knowing how to nourish growing salmon, they elected not to feed them; such starving and weakened hatchery youngsters likely served as easy prey for other aquatic organisms. Yet while Hokkaido's hatcheries did not increase adult fish numbers, they were still heralded as a great technological achievement. Every spring, visitors flocked to the grounds of Chitose Central Hatchery to picnic under the cherry trees that had been planted around the salmon ponds while celebrating the triumph of modern applied science. Multiple times, members of the Japanese royal family inspected the island's hatcheries, recognizing the facilities' work as an important national contribution.

However, hatcheries served Japanese state interests far better than salmon populations. From 1879 to 1893, the average catch of Hokkaido salmon was about seven million fish, with a peak of eleven million fish in 1889. Hatcheries seemed to be the perfect tool to supplement salmon populations subject to such intense fishing pressures, but they could not sustain such catches. Hokkaido's salmon stocks crashed. Despite increasing hatchery efforts, harvests hovered around three million fish per year from 1900 to 1970, less than a third of their late nineteenth-century levels (92). The problem was not that the Hokkaido fisheries managers were inept or improperly educated. Their results were no worse than those of US or Canadian fisheries professionals. Across the North Pacific, the hatchery technologies worked well to produce a modern material aesthetic but poorly to produce fish. Simultaneously, the suite of other frontier-making practices made it difficult to maintain salmon habitat in Hokkaido, as well as elsewhere around the Pacific Rim. The types of landscape modification needed to promote modern agricultural production completely altered the ecosystems with which salmon are intertwined. Within a few decades, dams, water diversions, industrial effluent, and sewage from urban areas rendered most of Hokkaido's major rivers unsuitable for natural salmon spawning. Industrial farm development denuded the forested stream banks that once provided shade to keep waters at the cool temperatures that young salmon require. Hokkaido developers' efforts to dike and drain riverine wetlands to prevent flooding and expand the land available for agriculture production virtually eliminated juvenile fish feeding areas, turning once meandering rivers into concrete ditches. Riverbed gravel dredging and pollution from paper mills and starch plants only added insult to other injuries. In short, by the early 1900s, Hokkaido's salmon rivers and their fish populations barely resembled those of a century earlier (Kobayashi 1980, 92–97).

Hokkaido colonial officials had immediately grasped that comparisons were technologies of landscape-making that could be harnessed for national and imperial development. For them, the materiality of comparative practices was self-evident; their enactments required the physical movement of bodies and technologies. Throughout the Meiji period and into the Taisho, Japanese exchange students, American and Danish advisors, cattle and plant breeds, and hatchery blueprints traveled in uneven flows in the process of bringing comparisons into being.

In Hokkaido, these comparisons were never reducible to "copying." Though Japanese officials used the American frontier as an example, they did not mindlessly reproduce its practices. Instead, they wanted to use comparisons with it to generate new configurations of humans and nonhumans in Japan—to use the power of comparison to create forms that would be at once legibly modern and distinctly Japanese. Although comparisons with the American West mattered greatly in Hokkaido's Meiji era development, the island was made through processes of creative and generative cosmopolitan thinking rather than through a single dyadic comparison. Because they wanted to make the island a symbol of modernization (*kindaika*) within projects of Japanese nation-making, officials were simultaneously making material comparisons between Hokkaido and Honshu, as well as between Hokkaido and Euro-America. The comparative aesthetic that has developed in Hokkaido—the mode of making similarity and difference—has not evolved from a lone binary comparison but through the negotiation of multiple comparisons at once.

The material legacies of these multiple comparisons sometimes unsettle visitors and residents. For many visitors and residents, Hokkaido feels uncannily "Japanese" and "un-Japanese" at the same time. Sometimes the uncanniness lies in small details like the decorative Japanese-Victorian moldings that linger under the eaves of Hakodate's buildings (Finn 1995). Sometimes it flashes up on a computer screen, as when one views the website of Hokkaido University, the direct descendent of Sapporo Agricultural College, which continues to cite "frontier spirit" as the first of its four basic educational philosophies. Still other times, it appears when one encounters the size and design of Hokkaido's farms, which are on average ten times that of mainland Japan and whose outbuildings more often resemble US midwestern-style barns and silos than Edo era stone storehouses (Iwama 2009, 2–9). One travel writer tried to explain this common sensation through comparisons with foreign lands: "In many ways, Hokkaido is the least 'Japanese' of all the main islands. It's Texas and Alaska rolled into one. It's Siberia. Switzerland. The last frontier and the end of Japan" (Ferguson 1998, 365). Many mainland Japanese and Hokkaido residents feel that the people who live in the north also march to the beat of a different drum; they are at once Japanese and different. In a newspaper article about Hokkaido, a Honshu man described northerners as people who live by different social codes: "When it comes to personal relationships [Hokkaido residents] are too easygoing. They're not interested in all the intricacies of status and hierarchy and just exactly how A relates to B. Without knowing these things, you just can't do business in Japan, and that's why Hokkaidoans lose out to mainlanders all the time" (Oka 1981). Such sentiments are widespread and often directly attributed to the island's frontier history, often in ways that erase Ainu people and celebrate settler colonialism. According to an article published in a peer-reviewed research journal, Hokkaido's "frontier spirit" has made the island's contemporary inhabitants more "psychologically" similar to Americans than to mainland Japanese (Kitayama et al. 2006).

As many people in Hokkaido will tell you, the island's frontier history has been embedded into it in ways that continue to shape the island's inhabitants, both human and nonhuman. Indeed, in a very literal way, places like the American West and the Columbia River are not external to Hokkaido but already materially within it. This is because practices of comparison have pulled pieces of these other landscapes *into* those of Hokkaido. It is productive to understand this island's landscapes—as well as, perhaps, many others—as sedimented layers of cross-cultural comparisons. When we take seriously such an idea, it requires that we study landscapes differently. It demands that we do not take landscapes as either isolated patches or "local" places. Rather, we must explore how landscapes are formed in relation to one another, often across large geographical spaces. We must ask how landscapes are tied together not only by commodity chains and resource extraction but also through heterogeneous practices of comparison. We must then trace the specificities of those comparative practices, querying how they may have made their way into the tissues of the world, into such materials as metal, plant fiber, and flesh. To examine how comparisons are embedded in Hokkaido's landscapes is to see an ostensibly "Japanese" landscape as itself cosmopolitan, as a place that is made by a set of comparative encounters in which comparisons continually bring other landscapes and cultures inside Hokkaido. This is a different sense of place, one that assumes that more-than-human landscapes emerge from routes as much as roots (Clifford 1997).

CHAPTER THREE

Of Dreams and Comparisons

Making Japanese Salmon Abroad

T HE practices of comparison that remade Hokkaido's lands and waters did not remain on the island or in the surrounding region. While Hokkaido was made into a site to experiment with bits and pieces of models borrowed from elsewhere, it also emerged as a ground for comparisons that moved outward from it to imagine how other places might be transformed in ways akin to the projects underway there. Hokkaido became a linchpin in several Japanese government and industrial initiatives across a wide swath of the twentieth century within prewar and wartime imperialism and postwar economic growth.

As we glimpsed in the Sapporo Agricultural College graduates' use of Hokkaido as a model for their later work in colonial administration and imperial governance, the island itself has served as a source from which to make comparisons that reach out and remake other natural and social land-scapes. As it trained many key colonial bureaucrats, SAC served as a prov-ing ground for logics and practices of Japanese expansionism that were applied within Japan's other imperial projects. At SAC, young Japanese men developed the comparative languages of civilization and backwardness along with the concrete skills in agriculture and scientific management that they deployed in Japanese colonial campaigns, and such know-how quickly became one of the island's most important exports, alongside goods such as canned fish.

Consider the example of Nitobe Inazō, an early SAC graduate mentioned in chapter 2, who became a key figure in the development of the colonial logics used to justify the Japanese occupation of Taiwan and Korea (Dud-den 2005). During his time as a personal assistant to Gotô Shinpei, the colo-nial governor-general of Taiwan, Nitobe transferred specific practices of agricultural management from Hokkaido to Taiwan. As historian Alex Dudden documents, one of Nitobe's projects in Taiwan was to introduce "large-scale sugar-planting techniques—a staple of global empire—that he had first learned during his courses with William Smith Clark, in Sapporo" (14). Nitobe was only one of many students who thought practices of empire through Hokkaido. In 1907, SAC became the site of Japan's first Department of the Study of Colonization and Agricultural Administration, and scores of the program's graduates went on to lead agricultural development projects in Japan's Asian colonies, with Hokkaido as a key point of comparative reference.[1] As Japanese expansion proceeded, Hokkaido's comparative importance continued. In the late 1930s and early 1940s, Japanese officials working in Manchuria's Reclamation Bureau turned to Hokkaido for inspiration for farming methods suitable to conditions on the continent, including cold weather and a shortage of labor. They soon invited experienced Hokkaido farmers to serve as advisors to immigrant settler communities and established still more experimental farms. In 1941–42, more than 250 Japanese immigrants to Manchuria were sent to farms in Hokkaido to serve six-month traineeships in the modern farming methods prior to dispatch (Tama 2012, 36).

While agricultural production took center stage, Hokkaido's salmon management practices were also drawn into Japanese imperial efforts. In the early twentieth century, the Japanese government constructed Hokkaido-style hatcheries across the Kurils and Sakhalin in an effort to mark and legitimate their claims to these islands and their fish.[2] The government also supported Japanese companies, such as Nichiro Gyogyō, in establishing salmon canneries in these new northern territories, inspired by the earlier success of Hokkaido's salmon industry. While these salmon projects ended abruptly when Russia reclaimed Sakhalin and the Kurils at the end of World War II (along with Japanese agricultural colonization projects in China, Taiwan, and other parts of Asia), Hokkaido-linked development imaginaries did not.

In the decades after World War II, Hokkaido gradually became central to a new set of comparisons. Reconfiguring imperialist patterns, this new wave of comparisons reenvisioned Hokkaido as a site from which to build practices of international cooperation, development aid, and fisheries supply chains, forms of global power that remained feasible after the war and the subsequent American occupation. One of these new-generation projects was an effort to transplant chum salmon and modern salmon hatchery techniques from Hokkaido to southern Chile. From the late 1960s to the mid-1980s, Hokkaido fish biologists, working under the auspices of what is now the Japan International Cooperation Agency (JICA), toiled to establish fish runs in Chile, a nation with no native salmon populations.[3] Despite the temporal gap, this project was linked to earlier imperial projects; its goal was to replace the salmon stocks that Japan had lost to Russia at the end of the war by creating new fish populations, specifically for Japanese consumption, in an overseas locale. The comparisons that this project's Japanese participants made between Hokkaido's and Chile's river hydrologies, infrastructures, and work rhythms also point toward the continued imperative they likely felt to perform their nation's modernity in the postwar period. When they brought Japan and Chile into the same frame, the fisheries biologists made comparisons that foregrounded the similarities of their landscapes (and thus the possibility of transplanting fish) while also emphasizing alleged differences in the countries' levels of technological and economic development (and thus the justification for Japanese involvement). After the end of Japanese empire but still working within dreams of Japan as a modern power, Japanese actors described themselves as creating a form of Chilean salmon production that would feed the economic growth of their more developed nation, while fiscally uplifting the Chileans in the process. The transfer of salmon stocks to Chile was to serve as a foundation for commodity-chain connections that would both funnel cheap fish to Japanese markets and help rural Chileans toward a higher stage of economic development. Within the legacies of earlier colonial comparisons, Hokkaido became a concrete site from which to jointly imagine how salmon populations might be established in new ecosystems and how postwar Japan might situate itself in the world.

The JICA-Chile salmon project, however, reveals comparative practices that are as idiosyncratic as they are structural. The comparative legacies of nation-state building certainly exert substantial force within the project, but they do so only within and through the complex braids of comparative practices that emerge within individual lives. In contrast to institutional logics of comparison, the comparative practices of an individual person reveal how particular practices of comparison emerge within biographies, as well as within state-making projects. The history of Japanese efforts to cultivate salmon abroad are evident in the unique comparisons of Aliaky Nagasawa, the fish biologist who headed the efforts to establish salmon populations in Chile that would mirror those of northern Japan. Nagasawa-san was indeed an agent of the Japanese state, and he undoubtedly took on many official sensibilities as his own. Yet he guided the salmon project by relying on his own eccentric and charismatic practices of comparisons, a heady and seemingly incongruous mix that combined Japanese nationalism with evangelical Christianity. The very unorthodoxy of Nagasawa-san's comparisons appear to have been integral to their world-making force as his intense passion for the JICA-Chile salmon project, which emerged at the unexpected intersections of his multiple comparisons, ultimately contributed a major spark to the Chilean aquaculture industry.[4]

When I first met Nagasawa-san in early 2010, about a year and a half before his death, it was clear that he had long accepted—even embraced—a spaciotemporal geometry in which the West equaled modernity and progress. Although he was in his early eighties, had a stiff leg, and carried a cane, Nagasawa-san had a distinctive walk that he attributed to a training program he had attended in Tokyo before the Japanese government sent him to Chile. The training program taught Japanese men how to walk more like Euro-Americans—upright, shoulders back, and with a bit of swagger to project confidence. At the training program, he also learned that it was more Western to wear one's hat slightly tilted to one side rather than squared stiffly front and center, and every time I met him, he wore his military-style felt cap cocked slightly to the left. Unlike most Japanese men, he also wore a mustache, a habit he picked up when he lived in South America. When Nagasawa-san spoke, his voice was rough from a lifetime of smoking, but his manner was gentle, and he gave conversations a cosmopolitan flair by peppering his mostly Japanese sentences with words of English and Spanish, languages that he had spoken nearly fluently in his younger days.

Nagasawa-san clearly relished fashioning himself as a sophisticated modern. After our interviews, we would often go out for dinner, and my suggestions of Japanese establishments were always rebuffed. At Nagasawa-san's insistence, we would invariably dine at an Italian or American restaurant, then spend the late evening listening to French chanson music at a European-styled café. As I drank red wine, Nagasawa-san chain-smoked long, slender, vanilla-flavored mini-cigars and discussed his passion for accordion music. After his wife had passed away, he had taken up the instrument and started music lessons, becoming such an aficionado that he made a pilgrimage to France a year before he died so that he could soak up *bal-musette* in its native environment. Nagasawa-san was also a Christian, something that he felt connected him to Western modes of being in the world. He had converted after he married his wife, a devout Protestant whom he adored. But Nagasawa-san's turn to Christianity was clearly not a token gesture to appease his wife. The process of conversion had transformed him; he described how he had accepted the Lord into his heart and come to experience the world through the lens of the Holy Bible, seeing nature—including salmon—as the work of God's hand.

In the 1960s, when Nagasawa-san was a young fisheries biologist at a Hokkaido fish hatchery, a group of Japanese fish processors had begun to worry about their increasingly limited access to North Pacific salmon. For nearly a century, Japanese fishing vessels harvested huge quantities of salmon in the North Pacific Ocean near Russia and Alaska. Between 1906 and 1945, under the terms of surrender negotiated at the end of the Russo-Japanese War, Japanese salmon fishermen ruled the Okhotsk Sea, filling their holds with fish intercepted on their return journeys to spawn in Russian rivers. But at the end of World War II, Japan lost control not only over Sakhalin and the Kuril Islands but also over its access to Russian-bound salmon. Although American occupation forces initially restricted Japanese fishermen to the areas around the nation's main islands, they soon relented, as part of policies aimed at alleviating food shortages in postwar Japan. With US general Douglas MacArthur's blessing, several fishing companies rapidly developed large salmon factory ships, which traveled across the North Pacific harvesting and processing salmon. But these ships soon raised the ire of American and Canadian salmon fishermen, who objected to the presence of huge Japanese vessels near their coasts, stealing what the North American fishermen saw as "their" salmon.

Drawing both international scorn and legislation, Japanese factory ship salmon harvests did not last long. Beginning with the Tri-partite Fisheries Treaty in 1952, continuing with the formal adoption of two hundred nautical mile exclusive economic zones in 1982, and ending with the Convention for the Conservation of Anadromous Stocks in the North Pacific Ocean in 1993, Japan's access to high-seas salmon fishing gradually disappeared as a new resource nationalism emerged. As fish tagging and tracking methods improved, salmon swimming in the open ocean ceased to be an undifferentiated mass of stateless creatures, a form of "nature's bounty" that was simply there for the taking. Instead, salmon became individuals who originated and belonged to a specific country, a country with specific rights to the salmon because the government had invested in their existence either by making them in hatcheries or by working to conserve salmon spawning rivers. Nations began to feel that they retained rights to "their" salmon even when the fish swam into extraterritorial waters. Under such logic, a US-born salmon swimming in Canadian waters was, at least conceptually, property of the United States. As a result of these ideological shifts, new international legal frameworks limited salmon fishing to coastal waters so that countries were more or less catching their "own" salmon as they returned to their rivers to spawn.

The Japanese government soon sought alternative sources of salmon. Japan's own salmon stocks—the majority of which were located in Hokkaido— remained depressed from decades of severe habitat degradation and over-harvesting. Developing Hokkaido had meant transforming its forests into farm fields and its rivers into irrigation and drainage ditches. By the 1960s, the situation was even worse: postwar make-work initiatives had included a number of river management and flood control projects in Hokkaido, fur-ther channelizing its waterways and lining their banks with concrete. In that same decade, the Japanese central government began to invest heavily in intensive and improved hatchery salmon production in Hokkaido, includ-ing through the funding of extensive research on juvenile salmon nutri-tion, management of diseases, and optimal hatchery release timings. But Japan's fish processors and distributors remained uneasy. For decades, Hokkaido's hatcheries had failed to bolster salmon numbers. Would this new generation of facilities be able to ramp up production and compensate for the loss of North Pacific fisheries? Or would it be better to expand outward once more? With Hokkaido's ability to produce large numbers of salmon as of yet unproved, industry members did not want to put all of their eggs (quite literally) into Hokkaido's hatchery baskets.

In the mid- to late 1960s, the Dai Nippon Suisan Kai, a trade industry group representing Japan's major fish processors, began to explore the pos-sibility of creating a new source of salmon beyond the borders of Japan. They had a wild idea: Why not create thriving salmon populations abroad that could be funneled to Japan through carefully crafted supply chains? As they imagined such a process, they began to compare Hokkaido's aquatic worlds to those of others around the world. Was there somewhere to which they might transplant Hokkaido salmon? Their first thought was New Zealand, a place where a handful of non-native salmon had already taken root. In the early twentieth century, New Zealand, which had no native salmon popula-tions, received crates of fertilized salmon eggs by steamship from California. Despite the odds, many of the eggs hatched, and their offspring were released into South Island rivers, where they established small, self-reproducing populations.

Based on such success, the Japanese fish processors thought investment in fish hatcheries in New Zealand could likely produce bumper crops of salmon that Japanese companies could then purchase. They contacted the New Zealand government to test the waters, but their offers of eggs and equipment were rebuffed. New Zealand did not need any more salmon, the government allegedly said. "They just had no interest in serious commercial fishing or salmon cultivation," a former JICA project member told me about the New Zealanders. "They had lots of sheep, so they didn't need salmon, and they just weren't that poor." With strong inheritances from the British sports-angler traditions, the New Zealand government saw the relationship between people and salmon as one of gentlemanly pleasure rather than commercial production. They were apparently not interested in a new paradigm, a potentially messy development project with significant risk in terms of environmental consequences and fisheries sovereignty.

After the New Zealand rejection, the Japanese fish processors needed a plan B. They had seen an American report that detailed some early efforts to transplant salmon to Chile. Although these efforts had not created self-reproducing populations of salmon as they had in New Zealand, they had had some modest success. A few juvenile salmon released into Chilean rivers had returned as adult salmon before funding, interest in the project, and the salmon runs themselves ultimately petered out. Based on such favorable information about the possibilities of salmon culture in Chile, the fish processor's group sent an exploratory party, which they referred to as a "mission," to Santiago, where their idea of creating a salmon industry was warmly received by the Chilean government. Chilean officials courted the Japanese processors and took them on a study tour of Patagonia, exploring possible sites for a Japan-sponsored salmon hatchery. After the mission, the Japanese processors reportedly prodded the Japanese government to create an official development aid project to establish salmon populations in Chile and coached Chilean officials about how to appeal to the Japanese government for such funds. The Chilean government soon submitted an application for aid, and the Japanese government responded enthusiastically.

JAPANESE DESIRES

The Japanese government likely embraced the idea because the salmon project meshed with its own dreams. Since the Meiji Restoration, resource scarcity has been a central concern for Japanese officials yearning to transform an island nation into an industrialized global power. Inspired by Great Britain, the Japanese government dreamed of an empire supported by resource-rich colonies. Fears of scarcity and dreams of imperial authority proved a toxic mix, driving Hokkaido colonization alongside twentieth-century Japanese military aggression and territorial expansion in Asia. They also led the Japanese government and Japanese businesses to consider a range of possibilities for accessing and extracting natural resources from Latin America. As early as 1889, a Japanese company established a joint venture mining business in Peru (Masterson and Funada-Classen 2004, 15).

Yet in contrast to mineral resources, which could be immediately put to use in Japan, many of Latin America's other products were not ready-made for Japanese extraction. When it came to agricultural products, there was a mismatch between existing Latin American goods and Japanese desires. As political scientist Toake Endoh explains, "The 'banana republics' served and had developed according to the interests of European colonial and U.S. capitalist interests. Latin America's traditional export goods—coffee, sugar, beef, and wheat—were not what Japan wanted. The Japanese preferred rice to bread, green tea to coffee, and seafood to beef" (Endoh 2009, 171). The Japanese government's solution to this problem was to send emigrants to Latin America to introduce and produce the goods that Japanese trading firms desired. They sought to create a Japanese diaspora that would produce commodities for the homeland.

Prior to the outbreak of the Pacific War in 1941, about 246,000 Japanese people migrated to Latin America (Manzenreiter 2017). Many emigrants were sent to Brazil and Peru as *kokusaku imin*, "immigrants under a strategic national policy" (Endoh 2009, 2).[5] With Japanese state funding, the immigrants were placed together in settlement colonies located in "undeveloped" frontier regions where the Japanese government urged them to undertake cultivation of the agricultural products most needed in Japan. Japan viewed these settlements "as an integral part of its colonization enterprise" and directly linked to expansionist policies in Southeast Asia (175). The Japanese Colonial Ministry coordinated activities among Japanese state-owned farms, farms owned by private Japanese companies, and independent Japanese farms that had been organized into agricultural cooperatives, and in the space of a few years, Japanese farms in Peru and Brazil began producing impressive amounts of cotton, pepper, and other agricultural commodities for export to Japan (175).

Although the end of World War II caused marked changes to Japanese practices of overseas resource extraction, there were some surprising continuities. New dreams of economic domination quickly took forms that echoed earlier dreams of territorial expansion. After World War II, Japanese concerns about inadequate resources only intensified as formal imperialism ended. In the immediate postwar moment, resource demand—of food, oil, and minerals—greatly outstripped domestic supplies, and the Japanese government turned from explicit colonialism to supply chains to move raw materials from extraterritorial hinterlands to the Japanese homeland. Initially, much activity centered on Southeast Asia, often in relation to timber extraction, with Japanese companies partnering with local elites (Dauvergne 1997).

In Latin America, Japanese efforts to maintain a resource diaspora also continued in the second half of the twentieth century, with an additional 93,405 Japanese citizens migrating to Brazil, Bolivia, Paraguay, and other South American countries between 1945 and 1989, often under bilaterally negotiated national contracts (Manzenreiter 2017). While languages of colonialism gave way to languages of "economic development," Japanese overseas initiatives in Latin America remained focused on bolstering Japanese agricultural and economic security. Even after state-sponsored Japanese emigration waned around 1970, the Japanese government persisted in its attempts to keep some of the continent's farm fields producing for its needs. The example of soy production in Brazil illustrates these new resource relations. After the United States issued a two-month ban on soy exports in 1973, Japan, a country heavily dependent on soy protein and unable to meet demands domestically, sought to increase soy production in Brazil, a country with little history of either soy production or consumption (Endoh 2009, 178). Through a combination of promotion by farmers of Japanese descent, direct investment by Japanese companies in soy plantations, and support from JICA, which provided funding and technical assistance for research on high-quality soybean varieties and management practices best adapted to Brazil, soybean production spread across the Latin American nation (Endoh 2009, 178, 232n23). Almost nonexistent in 1972, soy quickly became one of Brazil's top export crops (177). Japanese involvement in the development of the industry—especially through aid projects aimed at technology transfer and infrastructural development—networked the local producers of this new project with Japanese traders and markets. Extending far beyond soy, this coupling of international aid and supply-chain capitalism is emblematic of Japanese resource extraction.

As in earlier periods, the Japanese state continues to help secure the availability of such resources for Japanese traders, but such support now comes in the form of "development aid" rather than imperial decree. Japanese development projects have indeed emerged directly from the rubble of its imperial ones. In the years after World War II, "confronted by its own need for recovery and development, Japan invented a distinctive pattern of economic cooperation with the developing world that at its core is intended to contribute to Japan's own developmental plans" (Arase 2005, 5).

Japanese supply-chain capitalism and the cheap foreign resources that it helped acquire certainly aided the mercurial rise in Japanese economic power from the 1960s through the early 1990s. But the foreign-aid practices that were entangled with the production of such commodity chains were both a symbol of Japanese economic power and a method for building it.

From the 1960s onward, foreign development aid became an important way to enhance Japanese prestige. After two decades as a major recipient of post-war foreign aid (especially from the United States and the World Bank), Japan began to transition from receiving development aid to giving it (Takagi 1995). Such a shift was intended to be symbolic of a phoenix-like return of Japanese strength (Endoh 2009, 195). As Endoh (2009) highlights, in the case of Brazil, the Japanese received far more than a stable soybean supply from its investments:

> Another gain was in international clout. Japan's contribution to Brazil's economic development in the form of the formation of soy and related industries earned it credit in the international community. This was compatible with the values of postwar, peace-loving Japan in converting its economic power into international status and respect and becoming a superpower in development aid. (179)

Until the Japanese economy collapsed in the mid-1990s, Japanese government policies and transnational business ventures sketched a vision of an economically integrated Pacific that had many parallels with those found within earlier imperial imaginaries of a "Greater East Asia Co-prosperity Sphere." Since the mid-1990s, however, the Japanese government's ability to conjure its nation as a global economic powerhouse has dramatically declined. New mappings of Asia in which Japan's power is overshadowed by China have become dominant. In everyday conversations in Japan, people almost never cite imperial Britain or the postwar United States when they talk about their dreams for Japan's future; instead, they often express more modest dreams, such as the idea that Japan might emulate Finland or one of the other Nordic welfare states. But while Japanese dreams of geopolitical power appear to have waned, the supply chains that developed from them have not. The Japanese economy remains dependent on imports from abroad; to offer one example, Japanese domestic food production accounts for only about 37 percent of the nation's caloric needs (Statistics Japan 2020, 62). In an important way, ongoing trade in agricultural and natural resources between Latin America and Japan is not wholly different from the imperial practices that began with the colonization of Hokkaido, with its focus on raw materials for Japan's economic growth. The Japanese government's interest in salmon in Chile must be understood within the context of these unrelenting desires.

It would be a mistake, however, to focus only on Japanese desires. The particular comparative practices of the Japan-Chile salmon project arose not through the sui generis imaginings of Japanese participants but through the articulation of Japanese dreams with Chilean ones. Chilean government officials (both socialist and dictatorially inclined) yearned for stronger trade connections with Asia and a new export product that would help spur Chilean national development. Regional officials in Patagonia dreamed of a new industry that would make their area something more than an economic backwater. Wealthy Chilean sportsmen dreamed of home-grown trophy salmon that rivaled those of Europe.

Like Meiji era Japan, many nineteenth- and early twentieth-century Chileans also found themselves caught up in comparisons with Europe. Living in a former Spanish colony that had gained independence in 1844, elite Chileans of European descent often yearned for forms of nationhood, symbols of "civilization," and levels of development that would make them comparable to such powers as England, France, and Spain. Although such desires took many shapes, the oblong silvery bodies of fish were one of them. In northern Europe, salmon had long been considered the "fish of kings," seen as so valuable that their ownership was specifically mentioned in the Magna Carta (Montgomery 2003, 62). For hundreds of years, catching and eating salmon had been a pastime of the wealthy, and Europeans who moved to Chile sought to bring some of the high-collar civilization that salmon connoted to the New World. In 1865, coal baron Louis Cousiño stated that he wanted to transplant salmon and trout to Chile to create an angling paradise, but he died before he could act on such desires. Posthumously, his wife, Isidora Goyenechea de Cousiño, kept his dream alive; in the 1880s, she hired a Scottish fish expert, established Chile's first fish farm, and strove to introduce trout and salmon to Chile.[6] Successful rainbow, brown, and brook trout introductions soon followed, with both government officials and private citizens going to great lengths to transport these new species to Chile. For example, in 1905, the Chilean government ordered four hundred thousand Atlantic salmon and trout eggs to be purchased from a hatchery in Germany, shipped by boat to Buenos Aires in wooden boxes, sent by train to Mendoza, then carried over the Andes by mule to a Chilean hatchery on the Blanco River (Urrutia 2007, 457). Although many eggs died in transit, Chileans did not give up on their efforts to make landscapes that resembled those of Europe; by the mid-twentieth century, their perseverance had partially paid off. Although salmon had failed to take root, multiple species of trout could be found in most of Chile's lakes and rivers.

Yet at the same time that elite Chileans looked toward Europe, they also wanted to establish their own identity alongside economic independence. Part of this process entailed looking East. Despite its colonial connections to Europe—or perhaps, more properly, because of them—Chile, perched on the Pacific, has long been interested in fostering connections with Asia. In the late nineteenth century, the Chilean government made an important gesture of diplomatic friendship to Japan by giving the newly "open" country a warship, which Japan later used in the Russo-Japanese War.[7] By 1890, Chile had established a consulate in Tokyo, and in 1899, Chile opened its first Asia-Pacific embassy in Japan. From the start, the Chilean government hoped that transpacific trade would boost their nation's fortunes. Although exports to East Asia (metal and nitrates) were initially small, the Chilean government remained interested in such markets. During Japan's postwar economic boom—and after Pinochet's rise to power—Chile began aggressively marketing its exports across Asia. ProChile, the Chilean government's trade promotion arm, quickly established branches in Tokyo, Hong Kong, Bangkok, and Singapore. Such courting worked well in Japan; Chilean minerals and chemicals quickly became a part of its expanding economy, and with the addition of a growing trade in forestry, agricultural, and fisheries products in the 1980s, Japan became Chile's largest trading partner in the 1990s (Saavedra-Rivano 1993, 192–95).

These late twentieth-century desires to court Japan were part of broader Chilean governmental and industrial sector dreams of export-led development. Although actual development policies have changed along with Chilean governments, desires for "development" have remained constant. Although not an obvious candidate, since the 1960s, salmon have proved flexible enough to fit with nearly every Chilean administration's political aspirations and developmental dreams. During the administration of Eduardo Frei (1964–70) and the Christian Democratic party, salmon introduction initiatives were seen as an alluring possibility for assuaging the demands of fishermen who were protesting declines in harvests. Next, the salmon project fit with the goals of the socialist movement of Salvador Allende (1970–73), which focused on rural development and anticipated that salmon projects would create populations of high-value fish that could be harvested by local people in rural Patagonia (Winn and Kay 1974). For entirely different reasons, Augusto Pinochet's dictatorship (1974–90) was equally excited about salmon. Drawing on the expertise of the "Chicago Boys"—a group of Chilean economists who trained at the University of Chicago—the dictatorship implemented neoliberal policies that positioned privatization as the answer to struggles for modernization. The dictatorship saw salmon as a potential model for key neoliberal goals, such as the expansion of non-traditional exports and the introduction of external capital for development of private industry. Within this frame, salmon promised to be an exemplary tool for Chilean integration into international markets of goods, services, and capital (Urrutia 2007, 464). In its early years of rule in 1973–74, the military dictatorship enacted changes in Chilean law that were designed to help the kinds of industries that it imagined salmon might one day become, especially changes in export laws and tax laws that allowed for better competition in international markets. The idea of salmon production fit especially well with the military dictatorship's focus on the liberalization and privatization of Chile's natural resources. During the late 1970s and 1980s, while other Latin American nations moved toward resource nationalism and sought to limit the extraction of trees and minerals by international firms, "Chile went against the grain in Latin America by allowing foreign exploitation of its natural resources with few restrictions" (Saavedra-Rivano 1993, 202).

Salmon also meshed with elite Chileans' long-standing interests in species introductions in the name of economic development—and their deafness to such practices' ecological risks. Beginning in the mid-twentieth century, Chilean businessmen, sometimes with government support, imported animal and tree species that they thought might bring a profit. With hopes of creating a fur industry in Patagonia, South Americans imported Canadian beavers to Argentina in 1946 and American mink to Chile from 1930 to 1970 (Jaksic et al. 2002; Ogden 2021). However, the animals either escaped from their farms or were released into forests; feral beavers now destroy Chilean and Argentinian forests as mink munch on native species of rodents, terrestrial and aquatic birds, crustaceans, and insects (Choi 2008; Jaksic et al. 2002). When it comes to the plant kingdom, dreams of a lucrative forest-products industry inspired Chileans to introduce radiata pine (*Pinus radiata*) beginning in the 1950s. A scrubby tree in its native California, radiata pine grow rapidly when transplanted to other locales. With straight trunks and small, widely spaced branches, radiata pine seemed a perfect species for Chile's timber industry, easy to grow and easy to process. The radiata pine produced higher-value timber much more quickly than did native forests. After the Pinochet dictatorship initiated subsidized planting programs in 1974, these trees (along with eucalyptus) became the backbone of the private timber plantation system that blossomed in Chile. In south-central Chile, timber plantations increased from 29,213 hectares

(5.5 percent of the landscape) in 1975 to 224,716 hectares (42.4 percent of the landscape) in 2007 (Nahuelhual et al. 2012). Despite winning the praise of regional economists, these privately held monocrop forests have reduced species diversity and fragmented Chile's temperate forest habitats (Klubock 2014).

In this same spirit, elite Chileans, both in and outside of government, dreamed of salmon. Perhaps salmon could be Chile's *Pinus radiata* of the sea, a species designed to augment or even replace those of lower commercial value—in this case, the coastal shellfish harvested by poor and Indigenous communities. Yet salmon populations had already proven to be more difficult to establish than beavers, mink, or pine trees. Between 1870 and 1875, two Chileans, an entrepreneur and a scientist, partnered in the first attempt to bring Chinook salmon to their country, but their efforts ended in failure. In 1885, the Chilean government requested that a French veterinarian oversee another series of attempts to transplant salmon to South America, but the difficulties of transporting the fragile eggs across long distances again thwarted the project. Even after shipping improved, salmon introductions remained largely unsuccessful due to the complexities of their life cycle. Salmon typically migrate to the ocean then back to the river of their birth, but after transplantation projects released their precious progeny, the salmon tended to disappear into the ocean, never to be seen again. Such was the fate of the salmon who hatched from the 200,000 Chinook eggs, 114,000 sockeye eggs, and 225,000 coho eggs brought to Chile in the 1920s and 1930s from Alaska and the continental United States (Bluth 2003, 20). In the 1960s, the Chileans also formed cooperative salmon introduction programs with the US government, as well as with a private American company with ties to Union Carbide and Campbell's Soup (Borie 1981; Mendez 1982).

In short, by the time the Japanese mission showed up on Chilean shores in 1967 with their offer to transplant Hokkaido salmon, there was a lot already going on. Making salmon in Chile was a messy, multinational, multicontinent, multispecies affair in which Chilean agencies and business groups continually courted many parties, and investments in the early Chilean salmon industry were not limited to JICA.[8] These initiatives included the work of Fundación Chile, a nonprofit created through a partnership between the Chilean government and the US-based ITT Corporation, a manufacturing and communications multinational, which used its own funds to spur salmon industry development in parallel to (but in conversation with) the Japan-Chile project. Thus, a JICA project publication claimed that "the Chileans have long dreamed of salmon of their own"; it was not merely a platitude or justification for their own quasi-colonial desires to construct a salmon industry in Chile. Instead, the Japanese comment reflects how the JICA project intersected and developed within the flux of non-identical but overlapping dreams.

MAKING SALMON IN A WORLD OF DREAMS

Aliaky Nagasawa was working as a fish biologist in one of Hokkaido's salmon hatcheries in 1970 when a government official in Tokyo called him with a question: Would he be willing to try to create Hokkaido-style salmon runs in Chile? The moment that he said yes, Nagasawa-san found himself entangled in the complex web of Japanese and Chilean desires detailed above. But he was no newcomer to practices of imperialism and economic development, nor to their sometimes unexpected modes of comparison. Nagasawa-san was a child of the colonies himself. Although he had been born in Hokkaido in 1931, he was taken to Manchuria as an infant, not to return to Japan proper until the end of World War II, by which time he was already a first-year high school student. According to Nagasawa-san, the experience of growing up along one margin of Japan was what led him to work along another. For all its hardships, life in Manchuria was oddly cosmopolitan when compared to the intense wartime nationalism of mainland Japan. Although English language study was banned in Japan, Manchurian children were required to learn foreign languages (choosing among Chinese, Russian, or English). When he entered middle school, Nagasawa-san decided to focus on English, until the intensification of the war and its immediate aftermath upended his studies. Like many Manchurian families who survived the war and subsequent repatriation, the Nagasawa family decided to resettle near kin, specifically a sibling of Nagasawa-san's father who lived in the Hokkaido coal-mining town of Yubari. When Nagasawa-san reenrolled in high school, he found himself with a surprising educational advantage that would ultimately pull him into the salmon project. Under the US occupation, Japanese high schools were just beginning to require English language study, and although the tongue was entirely new to his peers, Nagasawa-san already had three years of instruction under his belt. With this head start, Nagasawa-san earned top honors in English throughout high school.

When it came time to apply for college, Nagasawa-san found the university entrance exams to be manageable, and he earned a seat in the Hokkaido University fisheries department. Fisheries science was a seemingly odd vocational choice for a man who had spent his life in inland China and a mountain coal town. But during Nagasawa-san's final year of high school, he had heard a lecture given by an older Yubari student who was attending a fisheries university in Tokyo. The student passionately claimed that someday, the coal in Yubari's mountains was going to run out, and he implored the students to turn their eyes to the renewable bounty of the ocean. Nagasawa-san, who already sensed that there was no future in the coal mines, was so moved by the speech that he decided to become a fisheries scientist. But coming from a working-class family, Nagasawa-san found it difficult to pay the bills for his studies. Fortunately, Nagasawa-san found that he was able to parley his relatively advanced English language skills into a part-time job at a nearby American military base. "It was pretty dirty English, all slang," he said of his time working on the base. "But it was English nonetheless. I got used to native pronunciation."[9] Those language skills would become unexpectedly critical to his future.

After college graduation, Nagasawa-san found work as a fish hatchery technician and researcher for the National Fisheries Service, and in his first few years of employment, he was stationed at several different Hokkaido hatcheries. Yet even when his rank in the fisheries agency was relatively low, he played an important role because he spoke the best English of any fisheries personnel. From his first year on the job, he was consistently selected to be the guide and translator for international guests to Hokkaido's hatcheries. When, as part of the initial phase of the Japan-Chile salmon project, the Japanese government extended an offer to provide technical training in salmon hatchery production to a Chilean in the late 1960s, they immediately contacted Nagasawa-san. At the time, there were no Chilean fish biologists who spoke Japanese and no Japanese fish biologists who spoke Spanish. But Rafael Aros, a young Chilean fisheries technician who had guided the Japanese mission from the Japan Fisheries Association during their visit to southern Chile in 1969–70, was eager to learn Japanese fish cultivation methods—and he spoke some English.[10] He was paired with Nagasawa-san, the fisheries person best able to communicate with him in that language.

When Nagasawa-san, by then the director of a small hatchery in rural northeast Hokkaido, was told that he was going to be assigned a Chilean trainee to mentor, he was surprised, but not shocked, to hear of a plan to introduce salmon to Chile. Several years before, in the mid-1960s, he had served as an official monitor and observer on a large North Pacific mother ship salmon vessel, ensuring that the private company that owned the vessel observed fishing treaties and regulations. In the evenings, he often dined and conversed with the boat's captain and other high-ranking staff, who were concerned that the increasing regulations, which Nagasawa-san was there to enforce, were likely going to put them out of business in the region.

"The people there knew I was in hatchery work," Nagasawa-san explained, "and they asked if it would be possible to make salmon somewhere else in the world. It was like a dream. But I said that it wasn't impossible, if you release fish somewhere in the South Pacific, there are lots of krill there, so if you let salmon go, the same kind of resources might develop there as what you have in the North Pacific.[11] It was talk about dreams [*yume no hanashi*]."

But these were serious dreams for people in the fishing industry, which is why, only a few years later, Nagasawa-san found himself tasked with teaching hatchery techniques to an earnest young Chilean. Aros, the Chilean, was even more surprised by the turn of events. Despite the historical interest in salmonids, when the Japanese mission arrived in Chile, "no one was thinking about salmon at all," Aros said. He had worked on freshwater fish culture to stock lakes for recreational fishing, but no one had thought about commercial production until the Japanese started searching for hatchery sites. Rather suddenly, Aros found himself entangled with the project and on a plane to Tokyo to spend half a year studying Japanese fish culture techniques. He was twenty-five years old, and although he had been over the border to Argentina and Bolivia, his experiences of international travel were limited. On his arrival, Japanese officials wanted to send him to study at a prestigious fisheries research institute in Osaka, but Aros knew that there were no salmon that far south in Japan. He insisted that he wanted to go north to Hokkaido, the heartland of Japan's salmon hatcheries. "But they said no, Hokkaido is too cold, you are from South America," Aros recalls. Fortunately, he had some photos of Patagonia with him. "When I showed them to the Japanese, they said, 'Oh, you have snow!' So then I was sent to Hokkaido."

After a stop in Sapporo to learn about the general structure of the Hokkaido hatchery system, Aros was assigned to study at a cluster of five rural hatcheries for which Nagasawa-san was the regional manager. "I was sent into his hands," Aros recalls. Despite their mutually imperfect English, the men worked together well. Aros learned about the differences among chum, pink, and sockeye salmon—how the chum preferred the locations in the river where springwater bubbled up through the gravel beds, how pink salmon populations fluctuate dramatically between even and odd years, and how sockeye salmon made long-distance migrations. He learned the procedures necessary for running a salmon hatchery—taking eggs, fertilizing them, hatching them, managing disease problems, feeding fry, and timing fish releases into rivers. Perhaps most importantly, Aros also learned things about work practices and scale that would serve him long after his formal participation in the JICA project ended. He was impressed by how the Japanese fisheries staff focused on efficient care of fish, not on human comfort.

"Americans wanted to heat hatcheries, but the Japanese did not heat buildings," Aros observed. "Who is the heat for? It is for the people not the fish." The scale of Japanese facilities also left a lasting impression on Aros. It made him realize the possibilities of large-scale fish cultivation. In the 1970s, most global fish production facilities were small-scale experiments with a few thousand fish, but in Japan, fish hatchery technicians had already pioneered processes for rearing millions of young salmon at a single site. In Hokkaido, Aros was able to learn techniques for fish production on a large scale, particularly how to rear high densities of salmon in small amounts of water by carefully managing the hydrodynamics so as to deliver oxygen to growing fish in the right moment in the right way. All these ideas would later become critical to the Chilean salmon industry.

When Aros finished his first round of studies in Japan in 1971, the Japanese government saw Nagasawa-san as a natural choice to send to Chile for the next phase of the project. Together, Aros, Japanese fisheries experts, and local Chilean workers were to construct a hatchery, rear chum salmon eggs shipped to Chile from Japan, and release the first chum salmon into Patagonian waters. According to Nagasawa-san, this is where the real story begins, what he considered the true adventures of salmon in Chile. "It's like a novel [roman mitai]," Nagasawa-san said of the salmon project's early phases. "Like tales of dreams [yume monogatari]." As passionate as he was about salmon, I think Nagasawa-san loved the stories of Chilean salmon just as much as the fish themselves. "It's really a dramatic story," he emphasized. "You know, there actually was a movement to make a TV drama about it at one point—a drama with real actors, not with me." Salmon, with their flashy silver sides, large leaps, and reliable returns to their natal streams, seem to court storytelling. For Nagasawa-san, the fish naturally hook people with their unique lives, reel them in with their stunning beauty, and refuse to let them go. "Salmon have a connection to the human heart," he told me. "They are fish that inspire feeling—the dream of these little fish going out and coming back big."

From Nagasawa-san's perspective, the ways these classic fish stories intersected with heavenly signs and human drama was what made Chilean salmon stories so epic. The project was about the unknown. At the time, the business of international aid seemed as much like terra incognita as rural southern Chile. The salmon project began not only before there was a JICA office in Chile but also before JICA even formally existed. The salmon project was not the product of a preexisting Japanese aid agency with agendas and plans but was instead one of the sites where the Japanese government experimented with what a formal international aid program might be.

Nagasawa-san said he felt like "007, James Bond. I was handed a slip of paper with a mission on it and that was it." In 1972, the Japanese government gave him three months' pay in cash and sent him off. Beyond the flight number and departure time of the airplane, he received virtually no instructions, no sense of how to proceed, and no inkling about conditions along the way. Perhaps he would receive such information once he arrived in Chile, he thought. When he disembarked from his plane, he was met by officials from the Japanese Embassy in Chile, who immediately took him to a Chilean government fisheries official. The Japanese Embassy man presented Nagasawa-san to the Chilean and said, "Here is the expert you requested," and "that was it."

The Chileans gave Nagasawa-san no more direction than did the Japanese. "I thought they would have requests or plans, but there was nothing," Nagasawa-san said. He thought he was coming to play a part in a grand Chilean plan, but the Chileans were expecting him to generate it. The day after Nagasawa-san arrived in Chile, a group of government officials convened a conference at which they began to grill Nagasawa-san about his plans for the joint project. He was caught completely off guard. "It seemed really rude to ask questions like 'What are you going to do for us? Why are you here?' to someone whom you'd requested." But the Chileans said that they couldn't make any plans because they didn't have the salmon fisheries experts to do so. The whole situation shocked Nagasawa-san: "In Japan, everything is always so top-down. People always give you a plan to follow, but here there was nothing. In Japan, the only people who make plans are the upper-level people in Tokyo and maybe the Hokkaido prefecture officials, but the concept of asking a technician like me to make a plan in Japan, it is just totally unthinkable."

Despite his status as a mere technician, he was the entire Japanese aid program in Chile, and he had to do something. Beginning in July 1972, he and Aros hastily drafted a plan with the knowledge they had, arranging for the construction of a fish hatchery and scheduling deliveries of equipment and salmon eggs from Japan for October through December. Because some Americans had already established a small experimental hatchery in the Los Lagos region, the Chilean government had asked the original Japanese mission to select a site farther south. Based on their ideas about what made for a good salmon river in Japan—cool, clean, well-oxygenated waters—they selected a location on the Claro River, near Coyhaique, a small town in the Aysen region.

Nagasawa-san described the area as "Hokkaido a hundred years ago." There were a few buildings with unreliable electricity, some radios, and a handful of cars, but otherwise not much. "Flying from Santiago to Coyhaique was like a time slip," he said. If he needed to send a message to Japan, he had to go Coyhaique's central phone office where it would take at least thirty minutes to get a connection to Japan. Because of long lines at the phone office, transmitting a short message to Tokyo could take all day. Although he was working with Aros, whom he knew from Japan, living and working in rural Chile was exhausting and lonely for a man who initially spoke no Spanish and who was worried about how to manage a major overseas project. "I didn't know any Spanish at all then, only *que será será*!" Nagasawa-san explained. Local residents seemed to have friendly feelings toward him, and people would stop him on the street to ask him to write their names in kanji, the Japanese script.

By November 1972, the future was beginning to look brighter for Nagasawa-san and the project. The hatchery was mostly built, the egg shipments from Hokkaido were in transit, and Yoshikazu Shiraishi, another Japanese fisheries biologist, had come to join the project. But then, just as it seemed the project was on track, tragedy struck. On the same day that the first Japanese salmon hatched in Chile, Shiraishi-san's heart began to beat irregularly. Although they called in a plane to transport him to a hospital in Santiago, he died of a heart attack on the way. Nagasawa-san tried to express both his grief about Shiraishi-san's death and his optimism about the salmon eggs in the short and simple telegraph he sent to Japan: "One side gone, other side born."

In the midst of the shock and grief, Nagasawa-san formed closer relationships with Aros and the other Chileans at the hatchery and lost himself in the technical dramas of making salmon. In their Coyhaique hatchery, Nagasawa-san and Aros began the difficult work of turning desires and dreams into fish flesh. The analogic comparisons between the climates, rivers, and oceans of Hokkaido and Chile did not offer tidy answers to everyday challenges; differences in hatchery rearing problems and post-release juvenile fish behavior seemed to trump similarities. Everything was trial and error, and Nagasawa-san said he felt more like an engineer than a teacher or expert. He had to invent ad hoc solutions without any advance knowledge or guiding theories. Nagasawa-san was not applying well-formed knowledge to a new locale, because at the time, people in Hokkaido and in the United States were still experimenting with salmon feeding practices and disease control methods themselves. And the applicability of the existing information was questionable. "It was all knowledge from the northern hemisphere," Nagasawa-san said. "Really, all of it was useless."

First, they had to deal with the difficulties of transporting salmon eggs from one out-of-the-way place (rural Hokkaido) to another (rural Chile). Logistically, purchasing Chinook or coho eggs from the United States would have been easier and cheaper than transporting chum eggs from Japan, but the use of American eggs was never seriously considered. "Politically, especially then, we had to use Japanese technology and materials," Nagasawa-san explained. "It was about nationalism in those days." The choice of chum eggs, however, was not only about nationalist dreams of creating Japanese salmon abroad; it was also about the specific migration pattern they hoped the transplanted salmon would take. In contrast to Chinook and coho, which stay relatively close to shore, chum make long-distance migrations in the open ocean, which Japanese boosters hoped might allow them to access more abundant food sources off Antarctica, ultimately supporting more robust populations.

However, transporting chum salmon to Japan was a monumental logistical task; because adult fish require large tanks and oxygenation systems, it was only feasible to move large numbers of fish when they were still eggs. To maximize survival, eggs were shipped as close to their hatching time as possible, when they were slightly less fragile, but with enough leeway that the eggs would not accidentally hatch during transport. Ideally, the eggs would spend three days in transit and hatch two to three days after they arrived. In general, sending the eggs from Japan to Santiago via Vancouver on Canadian Pacific Airlines worked acceptably. Once the eggs landed in Santiago, the Chilean Air force would speed the eggs to Coyhaique. Once, however, the Japanese government decided to ship some eggs via Frankfurt on Lufthansa; about half the eggs died when they got stuck on the tarmac during a second transfer in San Paulo, Brazil. Even under the best of conditions, things often went awry. On one occasion, Nagasawa-san opened a box of eggs only to find a sticky mess. Some of the eggs had hatched in transit, so there were live, dead, and dying eggs and alevins all jumbled together in what Nagasawa-san described as a grotesque "jam."

Tinkering with transportation schedules was only the beginning of their trials and tribulations. One of their major problems was the seasons—winter in Japan was summer in Chile. Salmon are only in egg form during the Japanese winter, which meant that shipments of salmon invariably arrived during the Chilean summer. Already stressed from their transoceanic and trans-equatorial journey, the young salmon—cold-water-loving fish adapted to short wintery days—were thrust into a world of long photoperiods and warm summer waters that they could barely tolerate. Normally, salmon would hatch in the winter and migrate to the ocean during the late spring, when creeks would fill with runoff from snowmelt. But a few months after their birth, the Chilean salmon faced a dry fall instead of spring floods. Nagasawa-san and Aros did not know what to expect. They released the juvenile fish that they had reared in the austral fall, but the fish just stayed in the river. As the fish grew larger and larger in the river, salmon project staff vacillated between hope and despair. They were relieved that the salmon were finding adequate food in the foreign river, but they worried that the fish might completely fail to migrate to the ocean. At last, in the austral spring—six months later than their counterparts in Japan—the large, well-fed young salmon swam to the sea.

But the salmon, a species known for their homing ability, were not following plans. "With such big fingerlings, we thought that we would have a high return percentage," Nagasawa hypothesized. "But we waited four years and no fish came back." At first, the salmon project staff thought that if they just raised the fish to a larger size before releasing them that they would be more likely to return, as larger, older fish tend to make shorter migrations. But although they kept releasing larger and larger fish, the salmon still did not return. Then, salmon project staff realized that brown trout, a species introduced from Europe, were eating many of the salmon as they tried to migrate down the rivers. When they examined the stomach contents of brown trout, they were filled with their carefully raised juvenile salmon. "We wanted to cry," Nagasawa-san said. "We wondered what we were doing. It felt like we were just releasing food for the brown trout." To address that problem, they began releasing juvenile salmon directly into the ocean, where they would not have to swim through a gauntlet of hungry brown trout in the lower reaches of the river.

But adult salmon still failed to return to the river. They tried different species of Pacific salmon, different diets, and different rearing strategies without results. Soon, they began to worry about the future of their project because of their dependence on imported eggs. At the time, salmon eggs were in rather short supply in Hokkaido, where salmon stocks had dwindled and hatcheries had little surplus; shipments of eggs to Chile were thus likely to be heavily curtailed, if not entirely terminated. So the Chile salmon project staff decided that they had to make their own salmon brood stock in Aysen to ensure that their program had a stable supply of eggs. "It's kind of odd for a Japanese to say this, but I could hardly wait to be independent from Japan," Nagasawa-san said. He also wanted his fish to be more "Chilean" than "Japanese." He felt that they had been doing it all wrong, trying to transplant highly developed eggs. Drawing on conceptions of citizenship based in natal location rather than blood, he felt that fish fully "born" in Chile would be better suited to that place than those that began their lives in Japan. He wanted to make juvenile salmon that were of Chile. He thought that they would do better in the new land if they had not known the scent of any other waters. Eggs fertilized in Chile would also be on the right seasonal cycle for the Southern Hemisphere, a significant advantage.

Nagasawa-san, however, had no expert advice to offer about raising adult brood stock. It was not done in Hokkaido hatcheries, as it was unnecessary. Each year, hatchery workers obtained brood stock from among the many fish that returned to the island's rivers. Without a tradition of adult salmon rearing in Hokkaido, JICA officials thought that penned salmon brood stock was a bad idea. "Why are you trying to teach things that aren't done in Japan?" they questioned. It seemed that the comparative logics that were to undergird the project had been stretched to their breaking point. But after Nagasawa-san insisted, JICA relented and went along with the proposal. Although the Coyhaique hatchery would continue to release most of its fish to migrate to the ocean, it would keep some in captivity and raise them to maturity so that the hatchery would have a supply of eggs that could not swim away. Because adult salmon crave saltwater, they installed a pen for the adults in a nearby fjord. Every aspect of the pen culture was novel for the team, but they somehow made it work.[12]

Yet even a steady supply of local eggs did not solve the hatchery team's problems. Chilean-born salmon still disappeared into the ocean, not to be seen again. The team considered still more explanations for their problems. Perhaps the salmon were surviving to spawn, but they were spawning in other rivers rather than returning to their home stream. Maybe the timing of the currents was not right and the salmon did not have time to get all the way back north to lay eggs. "Or maybe we just didn't have enough eggs, and we just didn't release enough fish," Nagasawa-san thought. "In that environment, if you get a 1 percent return, you'd be lucky." Finding a handful of surviving fish along the vast Chilean coast might be like searching for a needle in a haystack. As technological fixes failed, Nagasawa-san increasingly turned to his faith in God to make sense of the salmon. "If you think about it, it's really hard on the eggs and fry to transport them all the way to the tip of southern Chile. All you can do is leave it to the fish, to release them in this place and pray to God that they come back somewhere, anywhere. It just isn't about technological issues after that point. Just prayers for divine intervention [kamidanomi]."

Although the JICA project team was frustrated by the lack of returning adult salmon, they never doubted their goals. They firmly believed that salmon were good for both the economy and the soul. Nagasawa-san saw salmon as a fish of the global north, a literal embodiment of civilization (*bunmei*). Furthermore, as a Christian, multiplying the fishes to help the poor fit perfectly with his cosmologies, even if those fishes were ultimately destined to be exported to Japan. No one involved in the salmon project worried about the possible ecological consequences of introducing new species. "The idea that this was a non-native species that might damage the environment, nobody ever said anything about that," Nagasawa-san recalled. "Rather, everyone was interested in how the economy might become more active. . . . If you look at geological time, species have always been moving around." He felt he was creating a new ecology, a new salmon constellation, but he saw this as exciting rather than problematic. "In Hokkaido, it's bears and salmon, right? In Chile, it was flamingos and salmon." For his own part, Nagasawa-san also firmly believed that the southern hemisphere was at a lower stage of development, environmentally and culturally, and that this not only justified salmon introductions but made them a virtual necessity. He perceived extra room in the allegedly incomplete ecosystems of the Southern Hemisphere, and he believed that such space would allow new species to coexist, rather than displace, older species. Bringing salmon to Chile was part of finishing God's work: "Why did God not put salmon in the southern hemisphere? I guess he left that for humans to do."

While Nagasawa-san was committed to scientific methods, he also believed that the final phase of the salmon project could best be understood through languages of faith. As salmon failed to return to Coyhaique year after year, the project came under criticism from the Japanese government as a pie-in-the-sky project—a waste of money and time (despite its rather modest budget). But Nagasawa-san refused to abandon his faith. He believed, despite the lack of confirmed returns, that salmon were indeed swimming in the South Pacific. Like new Christians, the salmon needed time to grow in their faith, he said—in their case, faithfulness to a single river. They needed to go through a process of evolution and adaption. At other times, Nagasawa-san compared the salmon to the Israelites; they were living in diaspora and struggling to make their way in a new land. Overall, Nagasawa-san believed that God was testing his faith, much as God did to Job. In the context of that comparison, to give up on the salmon would be to give up on God.

During the eleventh hour of the salmon project, after the Japanese government had already decided to cancel it the following fiscal year, God finally spoke to Nagasawa-san through the salmon. In 1986, seven adult chum salmon were found in a river near Punta Arenas, far south of the project area (Shimura, Cardenas, and Nagasawa 1986, 17). The fish were healthy, mature, and robust, equivalent in size to the largest chum salmon in Hokkaido. For Nagasawa-san, the signs were unmistakable. Seven is the divine number of the Bible, he told me. Seven is the number of perfection and completion: the seven days of creation, the seven churches in Revelations, and the seven angels in the Gospel of Luke. Nagasawa-san saw the name of the river to which the fish returned as yet another mark of God's hand on the salmon project. The river was called Rio Ultima Esperanza— last hope river (*saigo no kibō*)—and the salmon's appearance there both brought a final sense of hope to the salmon project and reminded Nagasawa-san of the ultimate hope provided by God through the story of Jesus.

Even in 2011, the final year of his life, as Nagasawa-san faced seemingly endless suffering, including hospitalization for gastrointestinal problems, the death of a son, and a cancer diagnosis, he continued to believe in God and salmon, or perhaps God through salmon. He believed that the project would have been a huge success with bigger numbers of fish. "I really think it was possible. If we tried it again, I do think it would work. I would have liked to have tried it again, but the Japanese government was tired of it, and then the [Japanese economic] bubble burst and all." He fervently believed that the project had not been a failure, and that in some remote small river in southern Chile, there was an as-of-yet undiscovered population of chum salmon. For Nagasawa-san, the story was not over; the final chapter of the novel had yet to be written. There were still fish out there. One just had to believe.[13]

• • •

Nagasawa-san built webs of dreams through knots of comparison. His imaginative projects pulled him into comparisons, while his comparisons inspired his imagination. As a Christian, a fisheries scientist, and a man struggling to make a meaningful life, Nagasawa-san inhabited multiple sets of comparative practices. Indeed, it was his very mixing of modes of comparison—of talking about salmon as Israelites while relentlessly checking Chilean hatchery water temperatures against those in Hokkaido—that compelled him to fight to keep the Chilean salmon project alive and generated his charismatic rapport with the Chilean scientists he trained.

Overall, the JICA-Chile salmon project seemed worthwhile not only to Nagasawa-san but also, for many years, to large numbers of Japanese and Chilean officials, who saw practices of comparison that stressed similarities in biophysical parameters yet differences in stages in societal and economic development as commonsensical. When enacted in creative ways by various people within the national contexts of Chile and Japan, such comparative sensibilities generated nonidentical—but equivalently strong—desires for Chilean salmon. It is important to focus on individuals like Nagasawa-san because he was neither wholly idiosyncratic—dreaming up his comparisons alone—nor a comparative automaton, simply enacting established government logics. Instead, as he turned fuzzy project plans into actions on the ground, Nagasawa-san wove together—and ultimately reworked—the comparisons circulating around him, linking Hokkaido and Coyhaique as well as the registers of Japanese technology transfer and evangelical Christianity together in novel and surprising ways. While these comparisons may or may not have produced self-sustaining chum salmon runs in Chile, they have still had substantial world-making effects.

CHAPTER FOUR

The Success of Failed Comparisons

JICA and the Development of the Chilean Salmon Industry

I N March 2011, I stood on the metal deck of a salmon farm a few hundred meters off the coast of Chiloé Island. Wearing a bright orange life jacket and swaddled in a plastic gown, I had been required to disinfect my rubber boots and hands twice before I was allowed onto the floating platform. Without the Japan International Cooperation Agency (JICA) project, this farm would almost certainly not exist, and without my connection to Nagasawa-san, I would not have been allowed to visit it. During my travels in Chile, fisheries professionals had repeatedly warned me that I would likely never be allowed to visit a salmon farm because most facilities, worried about disease transmission in the aftermath of a fish virus outbreak, had barred visitors. Furthermore, as a young white woman, I was told that I would face the added burden of fitting the "Greenpeace profile" and was likely to be mistaken for an undercover radical environmentalist on a mission to discredit fish farms. Yet here I was, standing next to Alfredo Fuentes, the company's production manager dedicated to coho salmon farming. Standing on a scaffold of gently rocking walkways that surround a series of square net pens filled with juvenile coho salmon, Fuentes told me about the life cycle of these approximately three-inch-long young fish. Fuentes is a former JICA project member, and any friend of the project was clearly a friend of his. He was deeply grateful to the JICA project, which he credited with having given him the opportunity to know salmon and to transform the economies of southern Chile, as well as his own life. "Everything I know about salmon I learned from the JICA project," he said. "It has shaped everything for me."[1]

• • •

In the 1980s and 1990s, JICA's own reviews of the salmon project were luke-warm. Quite a few JICA officials saw the salmon project as an embarrass-ment, as a project that completely failed to achieve its technical goals of transplanting Japanese chum salmon to Chile. According to Nagasawa-san, "The people who authorized the funding for the project said, 'You spent that much money and don't have any results? It's over.'" On one hand, the JICA project and its most important comparisons did fail. The rivers of Chile and the water currents of the southern Pacific Ocean were not similar enough for Hokkaido chum salmon to thrive there. The kind of transplantation pro-gram that had successfully introduced Chinook salmon from California to New Zealand had proved unsuccessful in this context. But on the other hand, the JICA project turned out to be wildly productive—just not in the ways anticipated. Today, farmed salmon are big business in Chile; they are the county's number two export, behind only copper, and they generated about US$5 billion in 2018 (Salmon Chile n.d.). On the surface, this industry seems to have little connection to Hokkaido chum. It largely produces Atlantic salmon and rainbow trout, and it grows its fish to maturity in saltwater net pens rather than using the hatchery ranching systems promoted by JICA. Yet the JICA-Chile project has indeed contributed significantly to the for-mation of the farmed salmon industry in Chile's southern coastal regions.

The JICA project's failed comparisons—those that sought to transplant Hokkaido chum to Chile—successfully created a cohort of Chileans adept at both the technical skills for salmon cultivation and the cross-cultural know-how for building business relations with Japanese traders. These included Fuentes, with whom I stood on the deck of the salmon farm, and to whom Nagasawa-san had told me to reach out. A year before I traveled to Chile, as I sat with Nagasawa-san in a smoky Sapporo café, he showed me a Spanish-language magazine circa 2007 with an article that profiled the "Top Twelve" most influential people in Chilean salmon aquaculture. Among the portraits, Nagasawa-san pointed out the faces of six JICA project members who had received training at Hokkaido hatcheries. As the JICA project wound down in 1985–86, Fuentes, Rafael Aros (introduced in chapter 3) and the other Chileans involved with JICA did not give up on the silvery fish. Instead, they founded their own salmon enterprises, hired one another, and began building what would become a revolutionary farmed salmon indus-try. Through its technical training efforts, Aros and Fuentes insist, the JICA project indirectly provided a foundation for the larger Chilean salmon indus-try and for their personal successes within it.[2] Although the JICA project did not turn out as expected, the knowledge that the Chileans who partici-pated in it gained was key, as there was almost no salmon expertise in Latin

America in the 1970s and early 1980s. "Salmon ranching did not work economically, but it transferred knowledge," explained Aros, Nagasawa-san's closest collaborator, who went on to cofound one of the nation's largest farmed salmon companies. "We could take parts of that [knowledge] and apply it to farming. And it *worked*."

To succeed, the Chilean salmon industry needed to be able to produce healthy juvenile salmon that would thrive once they were put into saltwater net pens. The freshwater production techniques that they needed to make these juvenile fish were nearly identical to those that they learned under JICA's Japanese-style ranching system. At the JICA hatchery, as at all Japanese hatcheries, technicians must produce exceptionally strong juveniles if they are to have even a modest chance of surviving in the open ocean. "When you're making a smolt, you're making a fish that will live one year or more in another environment," explained Aros, "so the quality of the fish has to be very good so [it] can perform very well in those other conditions." Such knowledge about how to produce optimally healthy smolts was a major asset for Aros and others as they began producing juvenile fish for their pens. Much of the trick to producing robust juvenile fish, they knew, had to do with proper nutrition. Aros had extensively experimented with fish diets at JICA, and one of his most important insights was the importance of micropulverization and blending. Early fish foods were unsuccessful in large part because ingredients—such as fish oils, fish bones, and plant starches—were not well mixed. When tiny salmon took a bite of a fish pellet, they might be getting all fat or all protein depending on the part of the pellet that they munched. The grinding and mixing techniques were not evenly distributing the component parts of the feed, and as a result, the small fish were not getting the balance of nutrients that they needed to grow and thrive. By pioneering improved feed grinding and mixing techniques, the JICA project helped pave the way for more successful fish rearing. With such experiences under his belt, Aros cofounded a specialized feed production plant along with his fish farms when he entered the private salmon sector.

From the JICA project's efforts to maintain a few brood stock salmon to keep it supplied with eggs, Aros also knew how to raise adult salmon to sexual maturity and how to produce his own eggs for the next generation of farm-raised fish. While other early Chilean salmon farms were dependent on salmon eggs that they imported from Europe or North America, Aros's company could make their own. By using techniques learned from Japan and JICA to prevent fungus and other diseases, they ensured a ready supply of quality eggs without the costs of egg importation. Based on these kinds of Japanese-inspired hatchery practices, Aros and other former JICA project staff were able to make the most robust juvenile salmon of anyone in the fledgling salmon farm industry. Their production was so good that they began selling their extra smolts to other ocean-based salmon farms, which lacked the expertise to produce vigorous young fish in freshwater environments. Of course, such opportunities were a financial boon, but the benefits of their Japanese training went beyond their business balance sheets. "We developed self-confidence," Aros said.

The company Aros founded no longer uses practices that closely resemble those of Japanese hatcheries. The farmed salmon industry now rears Atlantic salmon, as well as coho and rainbow trout. The Atlantic salmon—much more delicate and easily frightened—have very different behaviors from their Pacific salmon relatives and thus require other techniques. As Aros explains:

> If you have coho in a pen, you take the feed and scatter it like
> you would with a chicken, and they jump out of the water to get
> the food. You do that with Atlantic, and they all run away over
> to the far side of the pen. When you have Atlantic, you have to
> have special automatic feeders. They have to be European
> automatic feeders, quiet, not noisy. You can use [Atlantic
> salmon techniques] with the coho, but the Pacific technique you
> can't do with the Atlantic [salmon]. So we mainly use the
> Atlantic system and put any kind of fish inside of it.

As a consequence of the mixed species production and the finicky nature of the Atlantic salmon, the company Aros founded has almost completely replaced Japanese technologies with Norwegian-based equipment and methods. But although the traces of Japanese influence on the industry have become increasingly hard to see, Aros continues to stress the critical role that the JICA project played. Even without the JICA project, the salmon industry probably would have developed in Chile eventually, he thinks, "but not with the strength that it did. [The JICA project] was very, very important. It was in the right moment."

The JICA project also gave the Chileans a leg up in building links to possible buyers. Making a new salmon industry required more than producing fish; it also necessitated the construction of supply chains and markets. After their grant-funded travels in Japan as the collaborative counterparts to the Japanese fisheries experts, Aros, Fuentes, and several other Chileans already had professional contacts, insights into Japanese business practices, and a respect for Japanese buyers' demands for high-quality products. According to Aros:

> The Japanese were confident that Chilean products would be
> good because the Chilean technicians had been trained in Japan
> and were using Japanese technologies. But more than that, this
> is something personal, something human, we had spent time in
> Japan. There had been a change in our minds, and we could see
> that [the Japanese] were very honest. It was impressive for
> us—to see the honesty. We could understand what they
> want. When the Japanese [buyers] came to Chile to deal, to
> begin buying salmon, the [other] Chileans complained, "Oh the
> Japanese, they always want something different, they always
> want something more." Well, we *knew* that they are perfection-
> ists. If you go to Japan and you buy something, it is good, it's
> perfect. . . . I took my wife to Japan three years ago. I wanted to
> show her the Japan that I knew. We went to hatcheries in
> Hokkaido. . . . My wife knew a lot of the Japanese experts that
> had worked here—more than twenty. And sometimes they
> complained about the quality of things [in Chile, saying], "Oh,
> nothing is good." Then she went to Japan, and said, "Oh, now I
> understand. Everything *is* perfect [there]."

Because Aros had been trained within the Japanese salmon system, his sense of an ideal salmon and ideal salmon farm practices largely matched those of the Japanese. "That made it easier for us to understand and be with the Japanese," Aros explained. "We thought that these are the right things. [For example,] we thought that the fish has to have this color." His company began importing rainbow trout from Sweden and Norway because those fish have the most silver-colored skin, a trait that is especially important to Japanese consumers. "The Japanese want fish that have that silver, which signals to them that it's not *hochare* [note Aros's use of a Japanese classificatory term].[3] You can have the best meat, but if the skin is discolored, it is second class. Because it is custom." Knowing this, Aros was able to improvise with the occasional batches of fish that had slightly darkened skin. "So if we have very good quality of flesh, but discolored skin, we take off the skin [and sell it to the Japanese that way]." Japanese traders looked favorably on Aros's company because it "followed their instructions." Even Aros himself saw the profound role that cultural encounters played in allowing his business to thrive, commenting that his stories were likely "interesting for anthropology."

Aros understood that the emergence of this salmon supply chain was dependent on embodied comparative practices shaped by particular cross-cultural exchanges. His very ability to imagine industrial production in Chile was made possible by his time in Japan and his capacity to think through Hokkaido hatcheries. Furthermore, his success with Japanese business partners was similarly enabled by his ability to maneuver in relation to the tacit understanding of Japanese fish preferences he had gained from his travels in Japan and his years of talking, working, and sharing meals with Japanese fisheries scientists and technicians. This was the ultimate success of the JICA project: it made people who could compare in new ways.

This new competency was transformative for the individual Chileans involved in the JICA project. With the exception of Aros, who had a prior university fisheries degree, the project gave skills to people who had little access to education, so the training had dramatic effects on their lives. In the early days of the salmon industry, people who had worked on the JICA project had such rare knowledge that farms allegedly paid them double the salary of other so-called experts. Although some of them are now retired, the majority of the JICA participants rose to high-level positions, becoming presidents of companies and heads of company divisions. For better or worse, the JICA project did not just produce a new industry whose profits were captured by existing elites; it actually produced new elites, in a process not altogether different from that of the Sapporo Agricultural Collage nearly a century earlier.

Alfredo Fuentes, whom we met briefly at the beginning of this chapter, is one of these new elites whose entire life was remade by the JICA project. He was a Coyhaique local with an education as an agricultural technician, a low-level position in the town's branch office of the agricultural ministry, and with limited prospects for career advancement when he was recruited to work on the salmon project. "The JICA project was my university," he said, and he managed to turn that training into wealth beyond his wildest dreams. When Fuentes first heard about the JICA project, he did not really know what to think about it, because he knew nothing at all about salmon. On one hand, he thought the project sounded "loco," but he also thought that if the Japanese were interested in it, then it was probably an idea with merit. He had faith in the Japanese.

When Aros invited him to join the JICA team, Fuentes had no idea how much the project would transform his life. Almost immediately, his respect for both Aros and Nagasawa-san deepened. "The JICA project was tremendously brave," Fuentes explained. It was so ambitious, and everyone approached it with so much dedication. The project twice sent him to study in Japan, an experience that was both professionally and personally transformational. Fuentes, a man whose informal, slang-filled speech style reveals a humble background, became a world traveler, a sushi aficionado, a small-plane pilot, and a valued salmon expert.

As the JICA project began to wind down in the mid-1980s, Fuentes, like the others, began to contemplate what he might do with his skills. He debated whether he should remain a civil servant. For him, the public sector offered stability but limited chances for advancement. "In the public sector, you can't really climb the ranks without a university education, but the private sector is more about skills," he thought. He had a good skill set for growing the emerging salmon industry, but new salmon ventures also entailed risk. Start-up businesses failed and companies merged, often leaving people suddenly without jobs. "In the public sector, they can't really fire you," he reasoned. But ultimately, the possibility of earning big money was too much of an allure, so he joined the other Chileans seeking to start their own commercial salmon ventures. Like Aros, Fuentes felt that he had a leg up on many of the other businesspeople who were also experimenting with commercial salmon farming in Chile at that time because he actually knew something about salmon from his years working with them at JICA. Although their commercial endeavor focused on coho salmon instead of chum, Fuentes found that "the basics are all the same, regardless of species." When Fuentes's farm began to produce marketable fish, it was able to build strong connections with Japanese buyers. When I visited Fuentes's office, it struck me that his basic Japanese, his familiarity with Japan, and the certificates of merit from JICA that adorned the office's walls all likely inspired a sense of trust and ease on the part of the Japanese traders who began purchasing his fish by the ton. Several years later, Fuentes's company was bought out by the larger salmon farm owned and managed by Aros and two other JICA project participants. Fortunately, Fuentes's fears of losing his job in the middle of a corporate merger were not fulfilled, and Aros made Fuentes one of his company's regional managers.

Today, the several Chiloé Island farming centers that Fuentes manages remain well known for their high-quality salmon (mostly coho), which they continue to sell to a predominately Japanese clientele. I sit in the back of his SUV as he drives at what seems like a maniacal speed down progressively smaller roads. As we head away from Castro, one of the main cities on Chiloé Island, the road is a paved thoroughfare, but by the time we near one of the fish farming centers, the road is a dirt track that seems likely impassable with a bit of rain. I am surprised by the poor road, imagining that a large salmon farm would require truck access for delivering juvenile salmon and hauling grown fish off to processing plants. But as Fuentes explains, only workers and occasional visitors use the road; all other materials arrive and depart by boat. When they harvest the salmon, they suck up the adult fish with a giant vacuum-like tube and take them alive by boat to one of their company's processing plants.

After passing through a metal gate, we arrive at an old European-style farmhouse on a hillock next to a tidal bay. The building, with its chipping blue paint, has been converted into an office, which is filled with a couple of computers, life jackets, and rubber boots. Some fifteen people work at the farm, but most of them are out on the platform rather than in the office. After getting outfitted with life jackets (for our safety) and plastic gowns (ostensibly as a sanitary measure), we tromp across the muddy tide flat where we stand and gaze out at the large metal grid that is the salmon farm. A small open motorboat suddenly speeds away from the salmon farm to meet us. When we hop aboard, we are instructed to step directly into a disinfecting footbath and to cleanse our hands with waterless alcohol sanitizer. In a few minutes, when we set foot on the platform itself, we are required to repeat the same hygienic procedure.

Fuentes comes here often to simply be with the fish for an hour or so. His office in Castro is where he takes care of recordkeeping, accounting, and other statistical management, but he spends much of his time on the move among the five farms he manages. "You can't grow a salmon sitting at your computer all day," he says. "You have to actually go out and look at the fish. The newer generation of salmon industry people just sits at their desks all day, and if something goes wrong with the fish, they just blame the computer." Through his JICA experience, he learned the value of "hands-on knowledge" gained from direct encounters with the salmon. When he spends time with the fish, he can draw on his instincts to detect problems with feeding regimes or disease long before they begin to affect fish growth statistics. When I prod him about the traits he looks for when judging fish health, he cannot explain it. He just senses it, he says. He just knows.

At the moment, Fuentes is checking out his 290-gram fish, which are set to be harvested in six months in November at between 2.5 and 2.8 kilos, net weight. The fish have come from the company's freshwater hatcheries in the Los Lagos region, where they were reared in metal troughs and net pens along the edges of freshwater lakes. Fuentes throws the salmon a scoopful of food pellets made primarily from meal derived from small fish, such as anchovies, mackerel, and sardines, produced by a plant a few miles down the road of which his company is a part owner. Thanks to careful management of the fish, he will ensure that most of them are between 2.5 and 2.7 kilos because that is precisely the size that Japanese chefs and housewives prefer. We stare at the clouds of fish in the mesh pens, ten meters square and twelve meters deep, each containing twenty-four to twenty-five thousand fish. As the fish grow, Fuentes will move some of them into thirty-meter by thirty-meter pens and reduce their farming densities.

Although the fish are still only the size of anchovies, their fate has already been decided. They will be headed, gutted, and frozen at a local processing plant, then transported across the Pacific to the Japanese buyers who signed a purchase contract for them about a month ago. Japan is Fuentes's top client and his top priority because they pay very good prices for high-quality products. But Japanese buyers are no easy sell, he says. "Dedication and hard work" are required to produce fish for them because the Japanese have "very exquisite tastes." Japanese buyers are "very attentive to *everything*," he says. In contrast to buyers of other nationalities, who often transact business by internet or in big city offices, those from Japan typically come to the site to see their fish in production. When the Japanese buyers are on-site, they look closely at the color of the skin, the color of the meat, and what kinds of medications the salmon farm is using, expressing clear preferences based on what sells in Japan. Japanese consumer preferences have indeed shaped Fuentes's production practices. For example, Fuentes eschews the twenty-four-hour artificial grow lights that are commonplace on salmon farms that produce for US and Brazilian markets. Although the artificial light speeds up fish growth, it also makes fishes' skin turn darker. While such a practice makes economic sense for salmon that will be filleted and skinned before they reach US restaurants and supermarkets, such a move does not pay off when selling to Japan, where, even though Aros may manage to sell some fish with its dark skin removed, most consumers still gravitate toward bright, silvery fish with intact skin.

Every time he receives a paycheck or sends off another load of fish to Japan, Fuentes is grateful to the JICA salmon project to which he feels he owes his personal financial success. He also praises the effects that the JICA project has had on both regional and national economies. Before salmon, there were few jobs in southern Chile, he says, and many people had to migrate to Argentina in search of work. Now, they can stay here, finding jobs at salmon farms, processing plants, fish feed factories, and other related industries. Salmon have also brought roads, airports, and improved water supplies to rural areas, including the island of Chiloé, where Fuentes lives, and have led to the dramatic growth of cities like Puerto Montt, where the population nearly doubled between 1992 and 2012, with another 12 percent growth from 2012 to 2017 (City Population 2021). On a larger scale, he says, salmon have helped diversify a nation that was too focused on mining and forestry. Fuentes sees the JICA salmon project as a critical part of Chile's economic progress. The JICA project, he says, was "a huge wake-up call" to the Chilean officials tasked with boosting development. Chileans knew that there were possibilities in seafood, he recalls, but not in this way. They were not thinking about cultivation at all until the Japanese began promoting the idea. From Fuentes's perspective, JICA's vision of salmon culture coupled with its investment in Chileans like him has left a powerful and positive legacy—especially in his own life.

CHILEAN SALMON IN JAPAN

One of the surprising dimensions of the Chilean salmon industry, however, is the degree to which Japanese demand for it needed to be actively cultivated. While fish farmers like Fuentes struggled to raise fish that would be appealing to Japanese palates, other Chileans were scrambling to spark interest in Chilean salmon among Japanese fish buyers and consumers. Today, it is hard to imagine that Japanese desires for Chilean salmon were not preexisting, as the fish are such a ubiquitous and naturalized part of Japan's seafood offerings. Based on my observations from 2007 to 2011, it appeared that even in Hokkaido, the majority of the salmon at supermarkets and sushi bars were imported from Chile. Such a phenomenon, however, was far from preordained; indeed, it seemed distinctly unlikely. In 1986, the first year that sizable amounts of Chilean salmon reached international markets, there were so many salmon from Alaska that the Chilean fish seemed unnecessary, one salmon trader told me. Although the JICA project had conjured a Japanese market hungry for South American salmon, that moment had passed. By the time Chilean salmon producers began making commercial quantities of fish, there was no longer a critical need for such salmon in Japan. Improved hatchery techniques had boosted domestic salmon harvests, and the booming Japanese economy meant that average Japanese families had no trouble paying top dollar for expensive sockeye salmon imported from Alaska. As a result, there was no ready-made market into which they could effortlessly slip. Chilean salmon farmers thus had to cultivate desire for their product as much as they had to cultivate salmon. And to do so, they had to tweak Japanese tastes.

During that first season in 1986, only a handful of Japanese traders had any interest in Chilean salmon products. ProChile, the government's trade promotion arm, hired Enrique Castañeda to expand this potentially lucrative, but seemingly difficult, salmon trade with Japan. Convincing Japanese traders to buy Chilean salmon seemed complicated, mysterious, and downright difficult. Castañeda's job, as he described it to me, was to open up the "black box" of Japan to the Chilean salmon trade. Although his background was in fisheries science rather than in business, Castañeda soon found himself in the role of promoter and cross-cultural negotiator. On the surface, selling fish to Japan would seem as easy as selling umbrellas in the middle of a sudden rain shower, since per capita, Japanese fish consumption is the highest in the world. But when Castañeda went to Tokyo to spread the word about Chilean salmon, the unfamiliar product received a lukewarm reception, one not altogether different from that faced by the first Hokkaido canned salmon in Europe. His fish were not only unneeded but also illegible. "Nobody understood about the salmon in Chile," Castañeda said. He encountered all kinds of confusing category problems. Before farmed salmon began to make their mark in Japan, *sake*, the Japanese word for "salmon," typically referred only to chum (*shirozake*), sockeye (*benizake*), and coho (*ginzake*), while other species, such as pink (*karafuto masu*) and Chinook (*masunosuke*), were grouped as "trout." Castañeda was marketing multiple species from Chile (including coho, Atlantic salmon, and steelhead/rainbow trout), all of which he saw as falling into a single generic category of "salmon," but he quickly found that category much less solid in Japan. Could farm-raised Atlantic salmon be sold as *sake*, or should it just be called *sāmon*, a Japanized version of its English name? Were steelhead *sake* or trout? "And the most difficult part," Castañeda said, "was explaining how we were producing 'Atlantic' salmon in the Pacific Ocean. 'No, no, no. . . . It is just a fantasy name,' [I explained]. It is the same species. It is the same fish." How could he make sushi shops comfortable with the idea of buying Atlantic *sāmon* at the fish market while selling it to their customers as *sake*?

When he arrived at the ProChile office in Tokyo, Castañeda spoke no Japanese, had no connections, and did not know what to do. So he began by phoning the Japanese Seafood Importers Association, which gave him a book with the names of all of the members of the association. Castañeda combed through the book and made a careful list of all the companies, big and small, that were dealing in salmon. Then he went to visit them in person, one by one. Although all the companies received him politely with a cup of green tea, only about four or five out of approximately two hundred companies showed any interest in Chilean salmon. But it was a start. Castañeda then began working with the ProChile office to organize salmon trade tours to Chile. "We paid for the tickets and selected and invited people. That started working," he said. Enticed by such free trips, more Japanese importers visited Chile, learned about its salmon, and became acquainted with Chilean salmon farmers.

But generating interest in Chilean salmon among Japanese salmon traders was only the beginning. The Japanese traders had to negotiate a market for the new Chilean products, which were noticeably different from other salmon at Japanese stores and restaurants. As we sit at his desk in a Chilean city, Shinji Aoki, a Chile-based Japanese salmon trader, pulled out his salmon color fan, resembling strips of paint samples, and pointed at a very pale pink color.[4] "At the beginning, it was like this," he said of the flesh color of Chilean salmon. "We were like, 'That's enough already. We don't need [such poor-quality salmon].'" Based on their experiences with Norwegian salmon, Japanese traders also had prejudices against farm-raised fish. "The first Atlantic salmon that entered Japan was from Norway, and the food pellets they use are different, I think. When you eat [that farm-raised salmon], it really stinks [of fish food] [*kusai n desu yo*]. It tastes bad, you know [*mazui n jya nai desu ka*]."[5] But in the late 1980s, Chilean salmon was so cheap that Aoki-san decided to take a gamble: "The first offer of Chilean salmon sold for a little less than three dollars a kilo. At the time, Alaska salmon was selling for two thousand yen per kilo [roughly $14]."[6] When we looked at the color we were like, 'We don't need it, but if it's only three dollars, well, I guess let's buy some.'"

To his pleasant surprise, in his opinion, the Chilean fish did not stink of fish food like those of Norway, something he attributes to Chile's high-quality fishmeal ("It just smells like *furikake*![7] Also, if you chew on the pellets, they aren't stinky or bad tasting"). But he had to figure out who might buy these new salmon, each species with its own traits. Coho flesh was so soft that it did not make good sashimi, because it was, in Aoki-san's words, *gucha-gucha*, or mushy. Coho did not work especially well in Japanized "Western" cuisine either, because when chefs took the skin off and the bones out to make fish easier to eat with a knife and fork, the coho meat would fall apart. But when grilled in the context of *katei ryōri* (Japanese homestyle cuisine), the bones and skin left on and the flesh firmed up by salting, the rich fatty flavor of the coho made it an appealing salmon choice. In contrast, both Atlantic salmon and trout-salmon are firm when raw, retaining their shape when sliced into sashimi and sushi and making them perfect for uncooked preparations. When he began importing Chilean salmon, Aoki-san was working for a Sapporo-based importer, and they quickly found that Japanese willingness to eat Chilean salmon varied geographically across Japan. For example, Aoki-san discovered that Hokkaido residents were willing to purchase Chilean-produced *trout-sāmon*, but they wanted nothing to do with the softer coho. "People in Hokkaido aren't afraid of new things," Aoki-san told me. "But they do know a lot about [seafood] quality." Because Aoki-san could not sell coho locally, his company began sending it to the Kanto and Tohoku regions of northern Honshu, where people liked the rich flavor so much that they did not care about the soft texture.

Over time, fish farmers, exporters, and importers all kept tinkering with words and equivalences in their attempts to build desires for Chilean farmed fish. While Chilean salmon might not be essential to Japanese fisheries markets in the sense that there were plenty of fish in the global market, Chilean salmon producers and Japanese importers worked hard to make their fish seem necessary. They needed to convince Japanese housewives and restaurant chefs that Chilean salmon matched perfectly with their emerging needs for cheap, healthy, and easy-to-prepare seafood. When it came to the allure of their low prices, Chilean fish had a stroke of good luck. In the early 1990s, the Japanese economy crashed. As the need to cut household expenses rapidly displaced desires for opulence, the charms of cheap farmed salmon began to draw consumers away from top-dollar Alaskan fish. But farmed salmon promoters did not just count on such historical conjunctures to create a market for them. Piggybacking on state-sponsored nutrition and diet programs, farmed salmon producers and traders promoted their fish as *kenkō ni ii* (good for health). In the past two decades, the Japanese government has encouraged increased consumption of omega-3 fatty acids, found in abundance in salmon flesh, and reduced intake of sodium. Japanese chum salmon, which have relatively soft, mild flesh, were typically firmed up and flavored through heavy salting, but the firmer and more flavorful farmed salmon are attractive even without salt.

Most importantly, however, farm-raised salmon boosters have promoted their product as *benri*, or convenient, for everyone. They are *benri* for wholesalers and supermarkets because they are available year-round rather than in a seasonal pulse. They are *benri* for convenience store *obento* lunch-box makers because farmed salmon can be made to order so that their fillet size fits perfectly into standard plastic trays. And above all, farmed salmon are *benri* in the kitchen. Traditionally, housewives bought and filleted whole salted salmon. But as an increasing number of women work outside the home and family life becomes more hectic, fewer people are interested in cooking labor-intensive food. In this context, many people in Japan told me that they see whole salmon as *mendokusai*, a bother or an annoyance. To prepare a meal with Chilean salmon, a wife need simply pick up a package of precut *trout-sāmon* sashimi from the grocery store and set it on the table next to the rice from an automated cooker. Added to curries, made into burgers, tossed into soups or chowders, breaded and fried, or simply grilled, farmed salmon can be served for breakfast, lunch, or dinner within the wide range of *washoku* (Japanese), *yōshoku* (Western), and category-bending recipes that are common in Japanese kitchens. Farmed salmon are also *benri* in that nearly everyone seems to like their taste. Because of their species, diet, and moment of harvest, they are generally oilier and richer in flavor than Japanese chum. This higher oil content tends to make them more appealing to young people raised on diets rich in foods such as meat and mayonnaise while remaining tasty to older people who prefer simple grilled fish and vegetables.

As a result of such conjunctures, by the early 2000s, farm-raised salmon had become so common in Japan that they came to define normative salmon. While people once yearned for the delicate taste and light pink flesh color of Hokkaido salmon, most Japanese now describe domestic salmon as dry and tasteless, preferring the fatty, bright-red flesh of pellet-fed and additive-dyed farmed fish. For the most part, Japanese consumers have literally swallowed such changes in salmon species and culinary practices without much thought. "Frankly, Hokkaido salmon just isn't that good," I once heard a Tokyo resident offhandedly comment. "The taste of farm-raised salmon is just better." Overall, Japanese consumers have become hooked on Chilean salmon. In 2010, salmon bested mackerel to claim the title of most commonly consumed seafood in Japan, a major accomplishment for a fish that did not even make the top five in 1965 (IntraFish Media 2010). Without the flood of Chilean fish, such salmon abundance would never have been possible.

RETHINKING "SUCCESS"

In the 2000s, JICA officials reassessed the Chile salmon project. In an about-face, they reversed their initial assessment of the project as a failure and heralded it as one of JICA's most illustrious achievements. After all, the JICA salmon project had incited real changes in Chile's economy and developed a solid new source from which Japanese consumers could obtain desired seafood products. Instead of interpreting the Chilean salmon industry as a fortuitous unintended consequence of a failed project, JICA officials began to narrate it as an outcome of smart project design. By emphasizing human capital development, technology transfer, and the formation of transnational business connections, the JICA project had allegedly built flexibility into its plans so that even if its original tack failed, its larger goals would succeed.

Indeed, the JICA project did a superb job of meeting the goal initially identified by the Japanese fish processors who prompted government aid investment in Chile: creating supply chains that would feed Japan with reliable and
cheap imported salmon. By the late 1990s, Chile had become the world's second largest salmon-producing nation after Norway, with Japan consistently
ranking as one of its most important markets.[8]

Regardless of its inability to directly produce fish, the JICA project linked
Japanese and Chilean salmon worlds in a profound way. By fostering human
connections, the JICA project ultimately played a significant part in linking
Japanese markets with Chilean salmon producers. Such connections were a
boon for people like Aros and Fuentes, but they were also important for Japanese buyers. The JICA project not only increased Chileans' familiarity with
Japanese salmon markets; it also increased Japanese familiarity with producers in Chile. JICA's involvement made Chile seem simultaneously less
risky and more accessible to Japanese companies and made Japanese involve-
ments in Chilean fisheries seem less threatening to people in Chile. In
short, JICA paved the way for corporate investment in Chile, alongside
salmon purchases. For example, aided by the rapport established by the JICA
project, Nichiro, a Japanese seafood giant, successfully created a Chile sub-
sidiary that established an early commercial salmon farm in 1981.[9]

As Chilean salmon have slipped into Japanese supermarkets, the Japanese government has increasingly trumpeted such successes. JICA has published a Japanese-language book that celebrates the salmon project as a
model endeavor, and more than one person described the JICA-trained Chileans who are still working in the salmon industry as JICA's "crowning
achievement" (gyōseki). In 2011, the emperor awarded Alejandro Aros the
prestigious Order of the Rising Sun, Gold Rays with Rosette, for his contributions "to the promotion of the technical cooperation of Japan in
Chile and the stabilization of food supply to Japan" both in his capacity as a
JICA project member and as the CEO of a major private salmon company.

But not everyone is convinced that these efforts to link Japan and Chile
through salmon have been an unmitigated success. Aliaky Nagasawa, for
one, was concerned that the industry had not benefited as many Chileans
as he had hoped. Although he remained a staunch supporter of Chilean
salmon, even Nagasawa-san sometimes gave its outcomes a mixed appraisal.
On one hand, he was deeply proud that he had helped empower Chileans to
start their own salmon businesses. When we shared meals, he enjoyed showing me pictures of the well-built streets, tidy sidewalks, and new buildings
of Chilean salmon industry towns, changes that he saw as both valuable and
linked to his contributions. But Nagasawa-san also regretted that rural

Chileans had not benefited quite as much as he had hoped. He was frustrated that high-quality salmon were not fully available in rural Chile, as the best fish are usually exported. In addition, he was dismayed that Chileans have increasingly ended up becoming laborers for salmon companies owned by Norway-based multinational companies rather than their own Chilean compatriots. This corporatization also worried him when it came to management practices; while he had faith in his trainees' abilities to make sound decisions about disease control and environmental issues, he was less optimistic about large-scale corporations where the managers who make decisions are too distant from the fish.

In Hokkaido, a number of people also began to worry about the effects of the Chilean salmon industry. Osamu Yamada is a retired hatchery manager who now teaches salmon education classes that try to persuade Japanese consumers to avoid farmed salmon. The classes, ostensibly for children, are equally targeted at their mothers, whose purchasing patterns Yamada-san hopes to change. Yamada-san begins his class by dissecting a large female salmon full of the roe that is considered a delicacy in Japan, encouraging the children to touch various fish organs and guess their physiological functions. Then Yamada-san turns his attention to the parents. Although he participated briefly in the JICA-Chile project, Yamada-san is strongly anti–farmed salmon. He sees imported Chilean salmon as essentially devil fish, fish who lead Japanese consumers to stray from the goodness of Hokkaido's salmon. In Yamada-san's talks, he explicitly seeks to reeducate Japanese taste buds that he sees as hijacked by the seductive and dangerous "other" of the fattier Chilean salmon. When Yamada-san asks the children to raise their hands if they like *sake*, the word that connotes domestic chum salmon, one child blurts out, "I like *sāmon*, not *sake*," meaning he likes imported farm-raised salmon but not domestic fish. "*That's* the problem," Yamada-san says. "[Japanese and Chilean salmon] are fundamentally different [*konponteki ni chigai ga arimasu*]." Chilean salmon are merely a product; they do not bring Japanese people into connection with their landscapes.[10] Japanese salmon, he says, also more properly nourish Japanese bodies. Yamada-san distributes a pamphlet that links the intelligence and high standardized test scores of Japanese children to their mothers' consumption of healthful fish. Such benefits, he says, are weaker for Chilean salmon, who, because of their pellet diet, have lower levels of beneficial nutrients such as omega-3s than Hokkaido fish. In contrast to the stories above about Chilean familiarity with Japanese protocols, Yamada-san describes Chilean salmon farmers as generic foreign producers, who, driven by profit motives, are unlikely to follow Japanese guidelines for producing a clean, safe product. Chilean salmon, as he describes them, are unnatural, contaminated with antibiotic residues and artificially colored.[11] Yamada-san aims to challenge the easy acceptance of farmed salmon by pushing people to reevaluate how they judge the *oishisa*, the tastiness of salmon. At the supermarket, farmed salmon, with their pretty red color, are *oishisō*, tasty looking. But such salmon present a false sense of *oishī*, he says. They taste good on the tongue but are not nourishing to bodies that would be better filled by Japanese chum. Yamada-san explains that despite their lighter color and less flavorful flesh, nutrient-rich Hokkaido salmon are the ones that are truly *oishī*.

Such critiques sit alongside other concerns about the ecological impacts of the Chilean salmon farming industry. While the exponential growth of salmon farms caught JICA officials, traders, and even fish farmers themselves by surprise, the outcomes of its proliferation were not entirely unforeseeable. As one Japanese fish trader told me,

> When I came to Chile, I thought that the industry would grow, but I never thought that it would become this big. I've also been to Asia and seen a lot of aquaculture there—shrimp, *unagi*. The final destination for all of these is always the same, I tell you. Increase the size of the industry, make a lot of money, there's a [fish or shellfish] illness, and then everyone runs away and the cycle starts over again. Unless we fix it, it is just going to be the same thing [with salmon in Chile].

The general problems of rapid aquacultural expansion have been made more acute by the specifics of southern Chile—particularly its relatively shallow bays and weak fisheries laws—which have exacerbated the spread of fish diseases. One can see the toxic conjunctures of geology and neoliberal economic policies in statistics about antibiotic use on salmon farms; despite its smaller total fish production, the Chilean salmon industry used almost 350 times more antibiotics in 2008 than did the Norwegian salmon industry, located in a region of deep fjords, strong tidal flushing, and stricter government oversight (Barrionuevo 2009).[12] Fish diseases and the chemicals used to treat them are only one of the ways salmon aquaculture affects its surroundings. As a result of the high densities of fish in small pens, fish excrement and uneaten fish food can sometimes become a pollution problem, killing nearby aquatic plants and shellfish on both ocean and lake bottoms. In addition, salmon who have escaped from the fish farms have become an invasive species in southern Chilean rivers, altering food webs, displacing native species, and changing watershed ecologies (Pascual and Ciancio 2007; Quinones et al. 2019).

As salmon farming has expanded in southern Chile, it has also remade the lives of its human residents. The industry has undoubtedly brought more cash to coastal communities, along with better roads, telecommunications, and other infrastructure (O'Ryan et al. 2010). Many Chileans living in salmon farming areas cite such benefits of the industry, along with increased local employment. However, not all residents feel that fish farming has been a turn for the better. Discharges from large-scale salmon production can damage the ecological assemblages of lakes and marine waters that have long been sites of artisanal fishing and shellfish collecting, while property concessions to salmon farms tend to enclose lacustrine and estuarine commons, further limiting local access to aquatic resources.[13] Such negative impacts—disproportionately located in rural areas with higher proportions of Mapuche Indigenous people—are not especially surprising. But as the following chapters explore in detail, Chilean salmon have also had less expected and frequently overlooked effects on the salmon worlds of Hokkaido—along with the comparisons entangled in them.

In the Shadow of Chilean Comparisons

Hokkaido Salmon Worlds Transformed

C OMMODITY chains connect producers and consumers, but they also link landscapes. Emergent from and with multiple strands of comparison, they transform tastes and purchasing practices, but they also alter multispecies arrangements. The Chile-Japan salmon trade is indeed a story about the relations and comparisons of hatchery technicians, international development officials, salmon farmers, wholesale buyers, and consumers. But it is also about how these comparisons have bound the salmon populations and biophysical conditions of watersheds in southern Chile and northern Japan into unexpected relations of coevolution and transformation.

When Chilean salmon, along with other farmed salmon from Europe, flooded markets from the late 1990s onward, they depressed global salmon prices.[1] However, these price declines were especially pronounced in Japan, where increased imports of Chilean salmon coincided with a rise in domestic harvests in Hokkaido to produce a glutted market (Shimizu 2005). Although Hokkaido salmon had been a scarce commodity for much of the twentieth century, by the 1990s, when Chilean salmon production began to take off, Hokkaido fish hatcheries had improved their practices and were generating a surfeit of fish. Yet because the comparatively cheaper Chilean farmed fish captured the eyes and taste buds of consumers, the price of Hokkaido fish also dropped, dramatically reconfiguring the island's fishing industry. Indeed, the prices for a portion of Hokkaido's salmon dropped so low that it did not pay to process them; while the roe from female fish—a product not produced in Chile—could still fetch acceptable prices, male fish, bearing only the flesh that competes against that of farmed salmon, were often left to rot on the docks, forcing the Hokkaido Federation of Fisheries

Cooperatives to take the unprecedented action of using its own financial reserves to buy them and process them into fishmeal.

Although such acute crises have abated, the upending of Hokkaido salmon markets by Chilean salmon continues to have rippling and surprising effects on wider land-water assemblages. While price declines have proved difficult for Japanese fishing communities, they have also opened up spaces for new salmon-human relations in northern Japan. In my initial fieldwork in Hokkaido, I sketched out a suite of new salmon management practices that had emerged in the 1990s and 2000s, often in comparative dialogue with and in counterpoint to North American conservation efforts: the fishermen who were protecting fish habitat to maintain salmon genetic diversity, the citizen groups that were trying to modify dams to aid fish passage and clean up polluted rivers so they could reintroduce naturally spawning salmon, and the earnest volunteers who were teaching schoolchildren to understand watershed ecologies. What became clear only later in my research was the role that Chilean salmon had played in instigating such changes. With the global glut of farmed fish, Japanese salmon were no longer viewed as a critical resource to be strictly managed by the state. The subsequent management shifts linked to such changes spawned new forms of "eco-friendly" fisheries initiatives, citizen-based conservation projects, and Indigenous-rights movements, with significant effects on the watersheds of northern Japan. By yoking together their salmon industries, the supply chain connections between Chile and Japan have simultaneously linked their watersheds in new ways.

In chapter 2, we saw how comparisons made in the name of Japanese nation-state development transformed Hokkaido's lands and waters throughout the late nineteenth and twentieth centuries. Yet there are other ways that comparisons have come to shape landscapes. As we saw in chapters 3 and 4, they are fundamentally intertwined with commodity chains, inseparable from experiments in extraction and production as well as consumer desires and market dynamics. They are embedded within the JICA-Chile project, the subsequent Chilean farmed salmon industry with its links to Japanese traders, and the cultivation of Japanese shoppers' yearnings for farmed salmon. As they are built into commodity chains that link projects of extraction, consumption, and accumulation, comparisons come to shape more-than-human landscapes in multiple ways.[2]

The farmed salmon trade that emerged from the historically contingent comparisons of people like Nagasawa-san and Aros played a central role not only by remaking the landscapes of southern Chile through direct impacts of farmed salmon production but also by creating conditions that have fostered surprising new practices of environmental conservation in northern Japan, with their own distinct practices of comparison. Like many other wealthy nations, Japan has an established history of commodity-chain extraction from distant lands.[3] Political scientist Peter Dauvergne (1997) has drawn particular attention to the environmental transformations that Japanese importation of raw materials wreaks on source countries, what he calls Japan's "shadow ecologies." Dauvergne traces how Japanese demands for wood products drive Southeast Asian forest exploitation. He shows that the supply chains that send Southeast Asian logs northward to Japan at a cheap price are "part of a complex process of interlocked indirect and proximate causes that drive unsustainable production and provide incentives and opportunities for illegal and destructive logging" in countries such as Indonesia (9). What has drawn less attention, however, are the ways that Japanese resource exploitation abroad also affects the ecologies that lie *within* the borders of Japan. While the production of farmed salmon for Japanese markets has cast an ecological shadow over southern Chile's coastal ecosystems, the reverberations of this process can be followed back to the landscapes of northern Japan. To borrow Dauvergne's language, how can we also see the "shadows" of his shadows, that is, the ricocheting effects of resource extraction abroad on Japanese more-than-human worlds?[4]

Tracing these connections is essential for understanding how comparisons and landscapes move and morph together. Doing so highlights how one set of comparative projects—those tied to the making of the Chilean salmon industry and its Japan-linked commodity chains—come to shift worlds in ways that foster new and distinct forms of comparison, with contrasting and divergent effects on more-than-human worlds. While the ecological effects of the farmed salmon industry in Chile confirm expectations about unequal exchange and environmental degradation, the effects of the Chilean farmed salmon industry in Hokkaido—and the comparisons emerging in their wake—are complex and surprising. As suggested above, by the time Chilean salmon finally reached Japanese stores in the late 1980s, the compelling reason for Japanese involvement in the industry—a lack of domestic salmon—had largely disappeared. The Hokkaido hatchery improvements in which the Japanese government began investing in the 1960s finally began to bear fruit, and by 1990, Hokkaido salmon populations had increased more than tenfold (Okamoto 2009).[5] Yet this bounty did not interest consumers, who gravitated toward the cheaper, fattier, and more brightly colored Chilean salmon that now sat alongside them in fish counter display cases. By the late 1990s, Japanese consumer purchases of chum salmon had dropped by more than a third in comparison with the early 1980s as salmon imports boomed (Criddle and Shimizu 2014, 288).

These Chilean farmed salmon imports dramatically changed the landscapes of salmon management in Hokkaido. During the food shortages that followed World War II, Hokkaido salmon were considered a scarce and critical nutritional resource, and the national government tightly regulated fishing while working hard to facilitate increased salmon production. During that period, the government literally inserted itself—in the form of metal weirs—into as many of Hokkaido's rivers as it could (Morita et al. 2006). The weirs blocked salmon from migrating upstream so that they could be funneled into further enlarged hatchery production schemes. Building on patterns set out in the late nineteenth century, this time with better technical success, the post–World War II government invested in hatcheries with explicit food security aims.

In the last quarter of the twentieth century, however, with the successes of domestic hatcheries and abundant imports from Chile, Hokkaido salmon have lost their status as a critical food resource. In response to this shift from scarcity to surplus—a surplus that continues today—the Japanese state has radically changed its relationship with salmon. The government has largely withdrawn from the work of making salmon, since there is no longer any reason to do so in the name of national food security. Instead, fishermen's groups are largely left to fund hatcheries and produce their own hatchery fish. Nearly all of Hokkaido's remaining hatcheries are now operated by private cooperatives, as salmon production has been reconceptualized as a business venture rather than as an essential state project. While some rivers still have weirs so that hatcheries can acquire their fish, these devices have been removed from a growing number of waterways.

By creating a large supply of salmon, the Chilean industry has helped create a space where Hokkaido salmon management can focus on something other than production logistics in the name of food security. As one retired Japanese fish hatchery technician told me, until the last two decades, with the concomitant rise in both Chilean salmon and Hokkaido hatchery fish, domestic salmon were "only food" for Japanese fisheries managers. They were not ecological beings or even biological creatures at all. "It was just 'let's increase, let's increase the salmon' [fuyasō]," the technician explained. "From today's perspective, it's hard to understand the concerns about food resources then. Now there's lots to eat, so salmon can be more than just food." Such abundance is creating new modes of conceptualizing salmon, he said. "Society is changing and salmon are becoming more biological [seibutsugakuteki]."

As he describes it, Hokkaido salmon are in the midst of a transformation from a mass-produced food resource into a wild animal with specific genetic and lifecycle traits that should be protected and conserved. Where Japanese hatchery production, like that of the North Pacific at large, was previously conceptualized as "sea ranching," with salmon positioned as the metaphorical cattle of the seas, they are now compared to regionally charismatic species, such as bears, Japanese cranes, and fireflies, as they are incorporated into frameworks of biodiversity alongside those of commercial value. Working within such paradigms, Japanese scientists and environmentalists are struggling to "un-domesticate" salmon by reducing reliance on hatchery reproduction, returning spawning to rivers, and reconnecting salmon to ecosystems.

In such ways, imports of Chilean salmon—which have set off a cascade of price declines, shifts in state-led salmon management, and reconceptualizations of salmon as wild animals—have significantly altered salmon management practices in Hokkaido. These changes are at once forcing and enabling people to take up new comparative practices, with wide-ranging consequences for both people and fish. These comparisons are emerging in settings such as a fishermen's cooperative, a set of salmon conservation efforts led by scientists and volunteers, and an Indigenous group's demands for comanagement. Chilean farmed salmon in no way predetermine these new Hokkaido salmon practices; however, they have so substantially shifted the conditions within which contemporary forms of Hokkaido salmon management come into being that it does not seem a stretch to state that without farmed salmon, these new sets of comparisons and human-salmon relations would not take the forms they do.

Overall, these twenty-first-century changes in the Hokkaido salmon industry seem to be largely positive, with greater commitments to ecologically oriented watershed management. But we cannot ignore that such transformations are coming into being in part through connections with Chile that have had less positive consequences for some of its people and ecologies. This point is not mere background but one that must be held in view when considering shifts in Hokkaido salmon worlds. Focusing on the increase of Japanese salmon conservation projects in the shadow of the farmed salmon trade also gestures toward a more general problem: how conservation projects in one place can be indirectly entangled with practices of environmental destruction in geographically distant locales. While conservation can certainly have positive benefits, environmental problems in sum are not always ameliorated, as production and extraction are often simply moving somewhere else.

Salmon both make large migrations across the North Pacific and remain intimately tied to the streams of their birth, to which they return to spawn. To date, salmon sustainability efforts have focused on improving the conditions of these spawning streams. Although biologically sound, activities such as planting trees, removing dams, and reducing point-source water pollution have led salmon managers to conceive of their work within frames of regional growth and local land-use planning. It may be time to reconceptualize salmon restoration within larger geopolitical frameworks and pose more difficult questions about what counts as conservation and sustainability. The lines of connection between the salmon worlds of Chile and Japan are not straight ones of linear causality but instead webs of ricocheting projects, diffracted and remade through multiple practices of comparison.

Existing and robust social science literature on transnational ties and commodity chains offers critically important conceptual resources for thinking about long-distance ties and, implicitly, the practices of comparison entangled with them, including those of corporate extraction, international development aid, traders, and consumer desires.[6] However, these approaches need to be expanded to better address the dynamics of the kind of linked landscape changes that appear in Japan-Chile salmon relations. To see these more-than-human aspects of comparisons and connections requires modes of attention that do not end at Japanese fish markets, grocery stores, or even taste buds but reach out into Hokkaido watersheds and its conjoined human-salmon worlds.

CHAPTER FIVE

Stuck with Salmon

Making Modern Comparisons with Fish

T HERE is no shortage of stereotypes about fishermen in Japan.[1] In the popular imagination, they are salty older men who speak with *hamaben*, a non-standard coastal dialect. They are assumed to have left school after ninth grade and to be more comfortable working with their hands than learning from books. Imagined as hard drinkers who live in weathered houses that dot the shoreline, they are supposed to be intimately tied to aging parochial villages, "vanishing" locales out of step with modern life (Ivy 1995). And perhaps most of all, fishermen are often described as *arai*—rough around the edges. I was living with Motozumi-san, a salmon harvester in the Hokkaido city of Kitahama, when his daughter's boyfriend was about to make his first visit. "I hear that her boyfriend is even more worried than normal because I'm a fisherman," laughed Motozumi-san.

Motozumi-san was laughing because he fits none of these fisherman stereotypes. He is in his early fifties, but thanks to the hair dye that camouflages his gray, Motozumi-san could easily pass as younger. Typically dressed in sweater vests, collared shirts, and khaki slacks, he looks professorial. He has two college degrees, one in business and a second in literature from a prestigious university. In his spare time, he reads Tolstoy and academic texts about the Roman Empire. Motozumi-san drives an expensive SUV that has not yet lost its new car smell, and his dinner table is a mix of imported Italian pasta and French jam alongside Hokkaido-grown white rice and whole milk from Japan's first certified organic dairy. In line with the fisherman stereotype, he does drink, but he prefers glasses of expensive Bordeaux over cheap beer. And he prides himself on his international travels. When I contacted him to check in after the March 2011 earthquake, he reported that he had missed my email because he had been vacationing in Australia.

When Motozumi-san talks about fishing, his words also defy stereotypes. He refuses the label of *ryōshi* (fisherman), instead referring to himself as a *gyogyōsha* (a fishing industry person) because he sees himself and his fisheries cooperative as producing a globally exported product rather than undertaking traditional harvest. Motozumi-san refers to his work as "business," using the English word, to connote its international legibility; at the same time, he fluently speaks the languages of macroeconomics and microbiology, describing how the price of the fish he harvests a few miles from his home are depressed by the production of farm-raised salmon in Chile, while regularly using concepts such as genetic diversity, nutrient cycling, and watershed conservation in a sophisticated way that was not out of place at the scientific conference I once watched him attend.

• • •

As we seek to understand salmon-human relations in Hokkaido, we must pay special attention to fishing cooperative members like Motozumi-san because they are the people who most directly manage the region's salmon populations. They are among the key people who not only act, but also decide how to act on salmon bodies, rivers, and coastal ecologies. In other parts of the world, including the United States, bureaucrats, scientists, and politicians exercise extensive control over the fisheries policies that shape day-to-day practices of hatching and harvesting salmon. Although US fisherpeople lobby for certain policies over others, their power to make their own management decisions is relatively circumscribed. State and federal agencies, not fisherpeople, do most of the work of monitoring fish stocks, restricting fish harvests, and implementing hatchery programs. Fishing in Japan, in contrast, is a largely self-regulated affair, with fisherpeople—not government officials—making the bulk of salmon management decisions. When I first began my research in Hokkaido, I went searching for the top-down national or prefectural salmon management policies (*gyogyōkanri seisaku*) that I thought must exist. But when I telephoned countless offices asking if they had any such policies, everyone seemed confused. "Fisheries policies?" they asked in puzzled voices. Finally, one official kindly explained to me that my search was in vain. Here, managing fish was the job of the fishers, he told me. "It's self-management [*jishuteki kanri*]. We give them advice, but there are no rules."[2] In the case of salmon, the Hokkaido prefectural government grants fishing rights to individuals and small groups and establishes a generous season during which salmon fishing is acceptable.

Beyond that, however, most management activities—including decisions about when to fish, how many to catch, how many to produce in hatcheries, and how to operate hatcheries—are the province of the fisherpeople themselves. Hokkaido salmon fishers, of course, do not make such decisions in an abstract space, divorced from the rest of their lives. Rather, their understandings of themselves and their worlds—their desires and fears, knowledges and lacunae—profoundly shape their management practices, as well as the structure of salmon populations themselves. In this chapter, I describe how the fisheries management approaches of salmon fishers in Kitahama, a city in northernmost Hokkaido, are intimately intertwined with their efforts to cultivate themselves as modern (*kindaiteki*) and international (*kokusaiteki*). In contrast to Nagasawa-san (chapter 3), who was pulled into a form of cosmopolitanism almost by accident through his colonial upbringing and his overseas job assignment, the self-titled "fishing industry professionals" of Kitahama sought to make themselves worldly moderns by design.

As the opening anecdote about Motozumi-san illustrates, Kitahama fishers are deeply passionate about cultivating cosmopolitan identities in which one's ability to compare well (i.e., to measure up favorably to others) is incumbent on one's ability to compare well (i.e., to make worldly comparisons). When I began fieldwork in Kitahama, I was thoroughly perplexed that the town's fishing industry professionals had almost nothing to say about fish. Instead, they wanted to talk for hours about their *kangaekata*, the "way of thinking" that they have used to build their lucrative fish-based business and worldly selves. As they described it, their *kangaekata* is at the core of both their "modern" identities and their "evolved" (*shinkashita*) fish management practices; as they describe it, how they think makes them who they are and shapes what they do.

Their *kangaekata* is a powerful practice of comparison where what matters most is one's very ability to compare. For the Kitahama fishers, one's ability to inhabit the world as a modern subject is incumbent on one's ability to make worldly comparisons. These fishers understand the world as composed of two kinds of people: those who can make such comparisons and those who cannot. As they seek to demonstrate the importance of comparison and make distinctions about who compares well, the fishers enact specific comparisons—between their fathers' generation and their own, between the *jidaiokure* (out-of-date) and the *kindaiteki* (modern), between the small-mindedness of the *inaka* (rural) and the *kokusaiteki* (international-mindedness) of the urban or foreign.[3] In their everyday lives, the Kitahama fishers link the ability to make good, knowledgeable comparisons to mobility, not in-depth place-based wisdom. People who are worldly and on the move can make cosmopolitan comparisons, while those who are stuck in place are parochial, traditional, out of date, and unable to compare. Rejecting the celebrations of local knowledges that their fathers embraced, they insisted that travel—often literal airplane flights—help equip them with the ability to perform flights of mind.

These fishers' practices of comparison are tied to their constant yearnings to enact what they see as modernity in an "out-of-the-way" place (Tsing 1993). Their modes of categorizing the world into the modern and out-of-date bring them into certain kinds of subjectivities vis-à-vis salmon, subjectivities that compel them to "rationalize" the salmon industry through particular notions of "rationalization" that they develop comparatively. In the midst of such practices, the Kitahama fishers reconfigure their relation to salmon, converting the fish from an emblem of local placemaking into a transnational commodity. Overall, this joint rationalization and commodification affects their relationships with and management of fish at the same time that it enables them to cultivate themselves.

A MARGINAL FISHING TOWN

Since the mid-twentieth century, "modern Japan" has become synonymous with its urban metropolises, with their bustling bodies, neon lights, and high-rise offices. In the postwar era, these industrial centers created economic opportunities that drew young Japanese to the cities, creating massive internal migration and rural depopulation. As cities bloomed, the countryside became cast as its outside; urban areas came to embody modern futures, while rural areas were depicted as "disappearing," with a mixture of romanticism and backwardness (Ivy 1995). As a result of Hokkaido's Meiji era colonization, the nostalgia that Hokkaido's rural towns evoke is less a nostalgia for traditional Japan than a nostalgia for dreams of strong economic growth and progress that the island's initial colonization conjured for ethnic Japanese—dreams that seem to have partially slipped away. Beyond the metropolitan area of Sapporo (Hokkaido's capital), the island's rural regions have had to cope with varying degrees of decline for much of the past half century. Beginning in the 1960s, Hokkaido's rural communities began to struggle as mine closures and agricultural mechanization decreased the number of local jobs, and an increasing number of rural Hokkaido youth, faced with bleak employment prospects, began to migrate to either Sapporo or south to Tokyo and other major Japanese cities.

Although located in the center of Hokkaido's most productive salmon fishing region, the city of Kitahama is caught up in these trends. Facing north toward Russia and the Okhotsk Sea, Kitahama is literally at the end of the line, about six hours by train from Sapporo. In Japan, train service conveys much about a place's ranking along the sliding scale of central to peripheral. In contrast to the epitomical bullet trains of Japan's busy commuter corridors that can travel at speeds of up to two hundred miles per hour, the train to the Okhotsk seacoast lumbers over mountain passes at less than thirty-five miles per hour. Because there is only a single track, the train must stop at a designated pull-off spot to allow the occasional train traveling in the opposite direction to pass. Inside the compartments, the seats are worn and the windows rattle. Unlike the quintessential image of the Tokyo metro trains so crowded during rush hour that white-gloved attendants push people into cars, trains to Kitahama are rarely full. Countless times, I have had the eerie experience during the last hour of the ride to Kitahama of being the only person remaining in my train compartment. Although people in rural Hokkaido most frequently travel by car, the presence or absence of train service still carries much symbolic value. Kitahama residents often told me with pride that unlike several other Okhotsk Sea fishing towns, they had not lost their rail service—yet.

Although the county-like zone of Kitahama has a population of about forty thousand, the city itself feels much smaller. Near the train station, there is a Kentucky Fried Chicken and a Pizza Hut combined into a single store, and a ten-minute walk down the road, there is a small strip of *izakaya* (pubs), a few sushi bars, and some *yakiniku* (grilled meat) joints. When I first went to Kitahama for preliminary summer research in 2007, there was a department store, but by the time I returned for longer fieldwork in 2009, it had closed. The town, economically sustained by a mix of fishing, farming, and tourism, is clearly not thriving, but neither is it in its death throes. At the same time that its downtown has nearly as many empty storefronts as it does stores, it also has a couple of new chain hotels and a sparkling hospital. Thanks to public-works monies, which also make up a substantial part of the local economy, Kitahama has a state-of-the-art public library, a community center, a concert hall, and two recently remodeled museums.

More than once, Tokyoites questioned my desire to spend time in Kitahama, a city that for them is synonymous with cold. Temperatures begin dipping below freezing in November and snow lingers as late as April. Kitahama's climate, and that of Hokkaido more generally, makes it seem temporally out of step with metropolitan Japan. In Tokyo and Kyoto, the cherry blossoms that mark the arrival of spring flower in late March, while Kitahama's buds do not open until May. As a result, many important community events, from elementary school sports meets to shrine festivals, are held on a different schedule in Hokkaido than in the rest of the nation to accommodate the weather. All this accentuates the feeling that northern Hokkaido, while Japanese, is also deviant in relation to normative Japanese-ness.

Fishing, one of region's most prominent sectors, is also an increasingly marginal occupation; only about one out of every 525 Japanese adults is employed in a job linked to the fishing industry (OCED 2021). Since the Meiji period, Japanese fishermen have consistently found themselves ensconced in an industry often viewed as less modern than other industrial projects. Japan's fisheries, rooted in collective sea tenure and hereditary rights transfer, are still sometimes seen as a "feudal remnant," as a holdover from "premodern" Tokugawa times.[4] While urban development and corporate innovation are seen as having brought Japan into the present, fishermen, who are seen as craftsmen, are understood as linking the nation to its past. Some Hokkaido fishermen embrace such narratives, which define them as "traditional" (*dentōteki*). For example, in Yamakawa, a southern Hokkaido town where I conducted participant observation, the fishermen loved to trumpet themselves as men of the sea, in line with the stereotypes that Motozumi-san bucks. Many of the Yamakawa fishermen were proud that they had started working in fisheries right out of middle school. When I asked them what they see as the most important trait for a fishermen, they almost all cited intuition (*kan)*. In concert with classic images of traditional fishermen, they see themselves as strongheaded, set in their ways and beliefs, and *wagamama*—egotistical, willful, and selfish.

The Yamakawa fishermen want to be "local." They sell the majority of their catch on contract to a single processing company just up the road, whose buyer shows up every morning with a medium-sized truck to haul the fish away. The fishermen also proudly make personal deliveries directly to nearby sushi restaurants, bars, and acquaintances, while their wives sell salmon, along with handmade seafood items, at a dockside stand. Their office exudes informality; the floor is filthy, and the tables, covered with scattered car magazines, have not been wiped clean. An old yellow fly-strip dotted with black insect bodies hangs from the ceiling, and a large nudie calendar featuring a big-breasted Japanese woman is tacked to the back wall. They scrape by financially, supplementing their fishing income with odd jobs, such as snow removal, during the winter off-season. Although the Yamakawa fishermen often wished for more money, they never expressed desires to be anywhere or anyone else.

In stark contrast, the Kitahama fishing industry professionals desperately want different lives and selves. They constantly chafed against assumptions about what kind of people fishers are, against the classic fishermen identities that the Yamakawa people embraced and embodied. While the Yamakawa salmon fishermen tended to speak in idioms of community, local products, and a sense of place, those in Kitahama did so in languages of professionalization, standardization, and internationalization. One morning, between the first and second waves of boat unloading, Motozumi-san and several of the other set-net group members decided that they wanted to switch their newspaper subscription to stay better abreast of current events. The office was receiving daily deliveries of the *Hokkaido Shinbun*, the major regional newspaper, but all the fishers gathered in the office already received that paper at home; here they wanted something different, something more focused on transnational political and economic issues to read during downtime at work. They decided that they wanted the *Nikkei* economic newspaper, the Japanese equivalent of the *Wall Street Journal*, which is usually read by businesspeople and college-educated professionals. Motozumi-san dialed the number of the newspaper distribution office to change their subscription. Although I could hear only Motozumi-san's side of the conversation, things initially seemed to go smoothly. He introduced himself as "Motozumi from the salmon set-net fishing group," and the newspaper distributor seemed happy to make a simple change from one paper to another. But a problem arose when Motozumi-san tried to explain which newspaper they wanted. "We'd like to switch to the *Nikkei*," he politely said. "No, not the *Nikkan*, the *Nikkei*," he clarified. But the newspaper distributor continued to assume that he wanted the *Nikkan*, a publication roughly equivalent to *Sports Illustrated*.[5] After a pause, he continued: "No, no, we *don't* want a sports newspaper. We want the *Nikkei*." Exasperated, he had to repeat his request several more times before the person on the other end of the line finally grasped his request. After hanging up, Motozumi-san turned to the rest of the office and commented about how the simple order change had proved rather difficult despite his clear pronunciation. "Even Heather-san understood me clearly, right? But that person just couldn't imagine that fishermen [*ryōshi*] would be reading the *Nikkei*!"

GETTING OUT OF KITAHAMA

The figure of the parochial fisherman that dogged the Kitahama fishing industry professionals was a part of their own pasts. Until recently, the lives and identities of people in Kitahama closely resembled those of Yamakawa fishermen. When Motozumi-san was a child, his father, Michio-san, was a quintessential fisherman. Born in Hokkaido, he had moved to Kitahama before the onset of World War II, where he harvested salmon as a laborer, not as a rights holder. During the immediate postwar sea reform, a redistribution of fishing rights from absentee owners to active fishers under the American occupation, Michio-san obtained his own salmon rights by joining with a group of men to form a set-net workers collective.[6] However, these rights did not lead to great wealth. During Motozumi-san's childhood, Michio-san's earnings were not enough to make ends meet, and he proved unable to provide for his wife and two sons. As a result, Motozumi-san's mother began operating a small drinking club for men (called *sunakku*, a cognate of the English *snack bar*) to make enough money to keep the family afloat. With his parents often absent, Motozumi-san was largely raised by a grandmotherly neighbor.

Motozumi-san and others of his generation did not want to follow in these footsteps. Growing up in an exciting postwar moment of increasing educational and economic opportunities in urban areas, they rejected the constraints of a lifetime of salmon fishing in Kitahama, a position they saw as both geographically and occupationally marginal. They did not want to be entangled in what they saw as suffocating structures of family legacy. Motozumi-san and a number of other young people from salmon families (mostly men but also a few women) managed to succeed in school and to make their way to good universities in Tokyo and Sapporo, despite the challenges of doing so from a rural area. Some were the relatively privileged children of Meiji era pioneer families who, in addition to their salmon rights, had significant accumulated wealth from colonization and other business ventures; others, like Motozumi-san, had only their own determination. Motozumi-san and others of his generation left town yearning to become "modern" by joining the massive urbanization movement. They dreamed of "making it" in life by making it *out* of Kitahama. Initially, their lives went as they hoped. Motozumi-san lived in Tokyo, worked as a journalist, and wrote a novel. Some of his peers became salary men, working for large corporations in several different cities. Still others found jobs through fish-related connections as buyers and sellers at Tokyo's famous Tsukiji fish market. One lived for years in England, while another worked in the office of a politician who later became prime minister. Such experiences and travels changed them. As they shifted locations—and moved away from Kitahama—the world seemingly opened up for them, and they became able to see and think in new ways. They were living their cosmopolitan dreams.

But although they ostensibly "made it" in Tokyo and cities abroad, one by one, the Kitahama youngsters felt compelled to return. They were mostly called home to deal with family matters, often ill parents or siblings in trouble. A few were lured home by ailing fathers who wanted them to take over their fishing rights and promised that since more and more fish were returning to the bay, lots of money was bound to follow. As one fisher told me, "Of course, I didn't want to return to Kitahama. I actually kept an apartment in Tokyo at first. My mother enticed me to come back in part by telling me that I'd make enough money in six months to live on for an entire year." But of course, things did not go as planned. One year stretched into several, and the promised good money from fishing ended up being so bad that the alleged off-season was spent driving taxis to try to pay the bills.

Although they felt duped, once the young men took over their fathers' fishing rights, it was difficult for them to quit. Because Hokkaido salmon fishing rights are hereditary, they were prohibited from just selling them off to someone else, as fishers can often do in other national contexts. Furthermore, once a family member gives up his rights, it is extraordinarily difficult—often impossible—to reclaim them.[7] Even though the fishing was far from spectacular, many of the sons who returned to Kitahama felt reluctant to let their rights lapse since they were part of their family inheritance, but they perceived this inheritance as a burden not a gift. They were stuck maintaining their families' fishing legacies until they could pass them on to another family member. The social worlds of Kitahama did not look favorably on children who forced their families to abrogate their rights. Gossip was prevalent, and young men and women who fled Kitahama after only a short time in the fisheries were criticized as "running away" (*nigeru*) from hard work, family, and community.

But after their time in Honshu's cities or overseas, the now worldly young men and women found Kitahama to be intolerably traditional, remote, and behind the times. They saw going back to Kitahama from Tokyo as just that: going backward. In the midst of their most modern of dreams, they suddenly found themselves entangled in classic filial stories of obligation and hereditary succession. While they yearned for routes, they got stuck with roots. Salmon fishing rights chained them legally and economically to a place they wanted to escape. Under Hokkaido Prefecture regulations aligned with the postwar fisheries sea tenure reforms, in order to maintain their families' rights, they had to make Kitahama their primary home, maintaining a permanent residence in the area where their net is located. The goal was to block the formation of a system in which absentee landlords—or in this case, absentee "sea lords"—owned net rights that local residents could work only as hired crew. If they had enough money, Kitahama fishers could own a second home in another city, but especially in the 1980s and early 1990s, when they had few funds, they could not get out of Kitahama. They described themselves as thoroughly stuck.

MAKING MODERNITY IN KITAHAMA

Through their experiences in Honshu and beyond, the Kitahama fishers had become certain about one thing: they now knew what "modern" living was supposed to be. Such knowledge initially heightened their depression about being stuck in Kitahama's fishing industry. But as they came to terms with the fact that they were not likely to leave Kitahama any time soon, they began to ask themselves about how they might create their own cosmopolitan identities in Kitahama. In the early 1990s, a core group of men, including Motozumi-san, decided that if they were stuck in Kitahama—and stuck with salmon—they might as well make the best of the situation. The younger college-educated Honshu returnees joined forces with a couple of established yet progressive fishers and started a conversation over beers and *shōchū*, a distilled beverage. What could they do to improve their lot? They soon formed what they called the Salmon Club, a coalition of fishers and local fish processing company leaders. The Salmon Club was a piscatorial consciousness-raising group, a gathering designed to develop what the fishers called *mondai ishiki* (problem awareness). The group was part of the men's attempt to see their financial problems as more than the inevitable fate of those dependent on boom-and-bust cycles and their social dilemmas as more than the inescapable consequence of having been born into a fishing family.

When they began assessing their problems, they initially focused on the large number of fishing rights holders in Kitahama. More than 160 people held salmon rights in the area, several times greater than in comparable areas of Hokkaido. How on earth could their modest fishery generate a decent income for so many? The large number of rights holders in Kitahama was something of a historical fluke. After World War II, fisheries throughout Japan—including Hokkaido salmon fishing—underwent phenomenal changes that aimed to democratize them. In a report about their fisheries reform efforts, American occupation officials wrote that they sought to take actions to "encourage the development within Japan of economic methods and institutions of a type that would contribute to the growth of peaceful and democratic forces," and that they sought "to favor policies which would permit wide distribution of income and ownership of the means of production and trade" (Hutchinson 1951, 6). Fishing was earmarked as an industry that had previously fostered acquiescence to authoritarian rule:

> The fishermen—those men who actually went to sea and caught fish—were virtually enslaved by the owners of ancient fishing rights which entitled them to the exclusive exploitation and benefits of the fisheries potentials within the area of the rights. Fisheries associations, dominated by government and/or local bosses, controlled the sale and distribution of the catch. The man who did the actual fishing was practically excluded from the benefits of his labor and was at the mercy of the controlling authorities without any chance of escaping from their grip or bettering his position. Far-reaching reforms of this antiquated structure were necessary to lead the industry into the ways of democratic organizations. (Hutchinson 1951, 6)

Across Japan, new laws established a fish cooperative system focused on developing principles of democratic self-governance. Cooperatives were to manage the resources within their assigned area, select their own members, distribute fishing rights to those members, and craft their own harvest regulations and rules for environmental protection.

In Hokkaido, the distribution of salmon rights took a special twist. Instead of granting salmon net rights to cooperatives to disburse to their members as they saw fit, Hokkaido Prefecture retained direct control over chum salmon, along with the island's limited runs of pink salmon and small numbers of trout.[8] In addition to joining their local fisheries cooperative, people who sought salmon rights had to apply through the prefectural government for the right to construct a net on a specific patch of sea floor. In contrast to other forms of fishing, which are usually undertaken with mobile gear such as nets, seines, or hook and line trolls, Japanese salmon are caught almost exclusively with *teichiami*, or fixed set-nets, also sometimes called pound nets or fish traps. At the beginning of each salmon fishing season, usually in August, salmon fishermen build set-net traps out of heavy nylon mesh, steel cables, and foam floats. These traps are precisely located along the seacoast so that migrating salmon, returning to Hokkaido's rivers, bump into their guide nets and eventually swim into their holding chambers.[9] Fishermen check and harvest salmon from these chambers by pulling them up on a regular basis—usually on the order of one to three days—emptying the

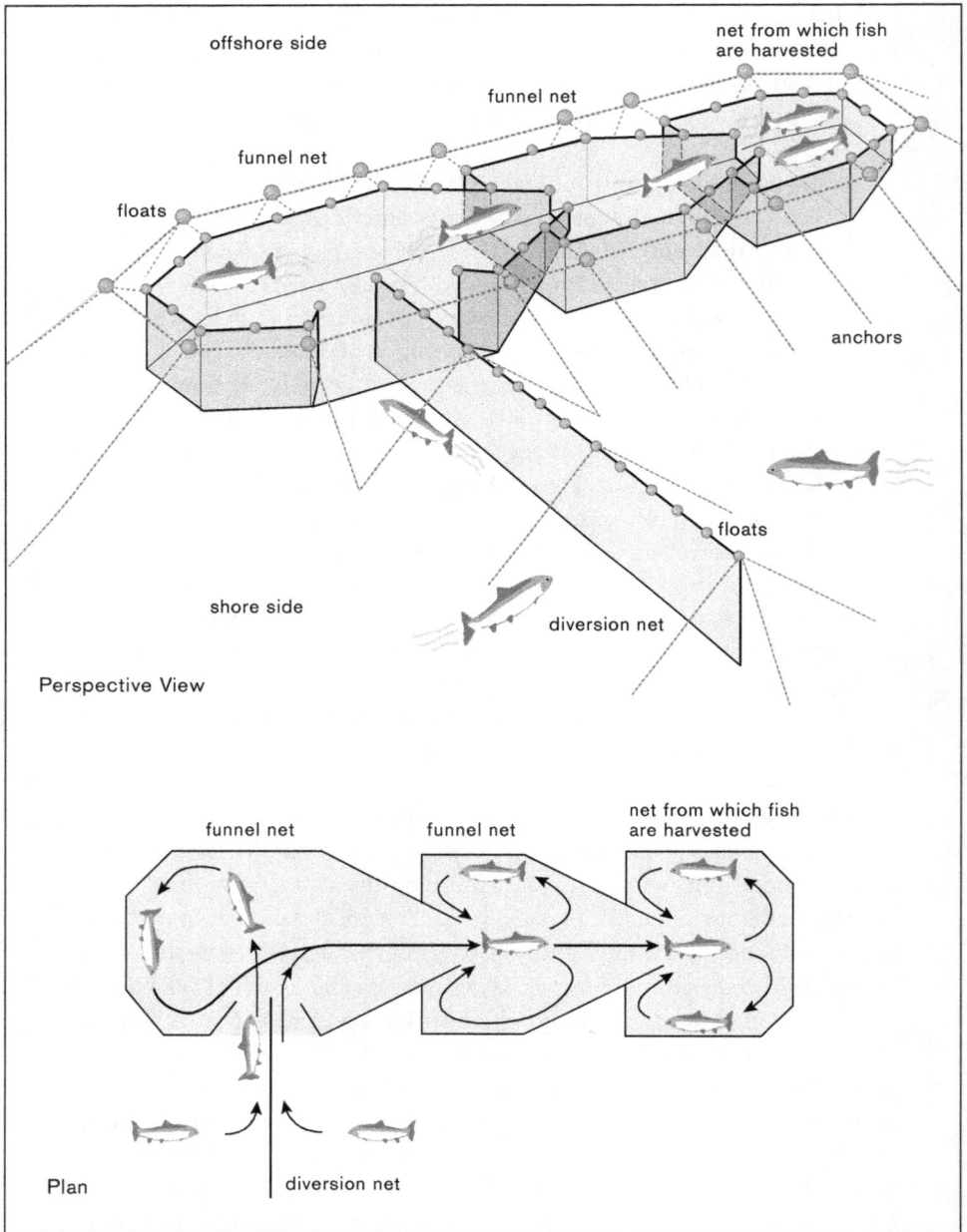

Labels in the figure (perspective view): offshore side; net from which fish are harvested; funnel net; funnel net; floats; anchors; floats; shore side; diversion net; Perspective View

Labels in the figure (plan view): funnel net; funnel net; net from which fish are harvested; Plan; diversion net

Diagram of the type of salmon *teichiami*, or set-net, used along Hokkaido's Okhotsk Sea Coast. The long guide net directs the fish into a series of chambers that funnel them into a holding pen, where they remain captive until they are removed by fishers. From a boat positioned alongside the holding pen, the fishers haul up the net, and dump the fish into the boat's hold. Diagram by Pease Press.

fish onto the decks or into the holds of their boats. Because of the large size and awkward shapes of salmon set-net traps, they have been entangled with different labor configurations than other modes of fishing. Where many coastal fishermen work alone or with a single partner, salmon set-nets require between seven and twenty people for their construction and harvest.

Because the nets are fixed, salmon fishing rights specify the size, shape, and patch of seafloor that each net is allowed to occupy. As a result of the specificities of salmon migration patterns, not all locations are equal, and during the postwar reassignment of fishing rights, people sought access to the most productive spots. In cases when there were competing applications for the same section of sea, Hokkaido Prefecture used a ranking system for determining who would receive the rights, with highest priority given to applications from workers' collectives, groups of seven to twenty fishermen who would work a single net together as owner-operators. If there were no such collectives, priority would then be given to smaller groups of fishermen who planned to incorporate. The lowest priority were applications from people who were seeking sole proprietorships. After the initial redistribution of rights in 1952, set-net contracts had to be renewed every five years, with owners retaining the right to renew their existing claim.

When fishermen applied for salmon rights in the immediate postwar period, the way the process played out on the ground varied by location. In many towns, the single-owner applications of prominent citizens went unchallenged, and salmon rights remained in a small number of hands. But in Kitahama, things unfolded differently. Across Hokkaido, prewar sole proprietors of salmon rights had hired migrant laborers from Honshu to haul in their heavy set-nets. In most places, the laborers stayed in Hokkaido only during the autumn and early winter fishing season, keeping their Honshu villages as their primary home. But in the case of Kitahama, a sizable percentage of salmon laborers had permanently relocated to the city. Thus, when a chance at fishing rights arose in the postwar scramble, these laborers were legal community residents who wanted their share. Furthermore, the postwar period brought an influx of skilled fishermen to Kitahama, as many returnees from Japanese settlements in Sakhalin and the Kuril Islands ended up resettling along the Okhotsk seacoast. With so many fishermen in town, people in Kitahama were forced to form *seisan kumiai* (workers' collectives) or at least other forms of joint ownership in order to have a chance at securing rights to a net under the preference system, resulting in a fishery whose proceeds were divided into a large number of small portions, leaving most people financially strapped.

The Salmon Club members also identified other structural problems with their salmon fisheries. While their fathers had struggled with a lack of fish, by the early 1990s they had begun to struggle with too many of them, both in Kitahama and in global salmon markets. Just after World War II, when American occupation officials conducted a survey of Hokkaido's fisheries resources, there were so few salmon remaining that they did not even seem worth counting. While the survey specifically listed the number of harvested tons for the most commonly caught species such as herring and squid, salmon were simply tallied under the category of "other fish."[10] In the 1970s and 1980s, however, improvements in hatchery technology and favorable ocean conditions caused a nearly tenfold increase in salmon numbers, transforming northern Hokkaido's hatcheries, which had failed to boost salmon numbers in the previous hundred years, into fish-making machines (Okamoto 2009).

But this dramatic increase in salmon did not solve the woes of the Kitahama fisherpeople. As discussed in the interlude, in the 1990s, Chilean farmed salmon began to flood Japanese markets, resulting in significant price declines within Japan and on global markets. Although a profitable domestic market remained for a handful of the highest-quality Japanese salmon, the Kitahama fisherpeople were routinely forced to sell the majority of their lower-grade fish at rock-bottom prices, often to fishmeal or fertilizer companies. As their fathers had been when salmon runs were weak, the new generation of Kitahama fishermen remained poor—this time a kind of poor they called *tairyō binbo*, or "big harvest poor." On top of sluggish markets and too many fish, they also struggled with uncertainty. Based on the slightest differences in water temperature and currents, the routes that the salmon took through Kitahama Bay varied from year to year, as did the specific nets they entered. In a given season, the salmon flooded some nets, while others stood almost empty. Because individual fisherpeople held rights to only a part of one net (or to parts of a handful of nets in the same area), their earnings swung dramatically from year to year, and a bad season could be exceptionally tough for those with little savings in the bank.

MAKING INTERVENTIONS

What then were Motozumi-san and the other young Kitahama fishers to do? They might be stuck with salmon, but they decided that they were not stuck with this form of fishery and its problems. The trials of local fishing were not inherent, they asserted, but the product of parochial and backward thinking. A good life was possible if they could manage to overcome old thought patterns and ingrained practices. For them, a good life meant many things. It meant having enough money to send their children to college, fly to Sapporo on weekends, take overseas vacations, and pursue hobbies, instead of menial part-time jobs, during the off-season. It meant a world that privileged hard work over seniority. And it meant having a spotlessly clean business office filled with computers and spreadsheets rather than a grimy bunkhouse. To bring such dreams into being, they felt they needed to work together in new ways. In other locales, one might be able to fashion oneself as a self-made cosmopolitan, but in Kitahama, modern identities were going to require collective effort.

The Kitahama fishers decided that they needed to begin by reforming their organizational structures. The sky-high costs of salmon fishing that each set-net group bore were consuming most of their potential profits. Each net group was an independent business unit with its own office, office staff, storage area, and shop building. Each also owned its own boats, nets, and other gear. Furthermore, because the number of rights' holders was often less than the number of people required to haul in a net, most set-net groups hired migrant laborers from northern Honshu to help them during harvest time, so they also maintained their own residential bunkhouses, providing room and board in addition to salaries. Once all these costs were paid, there was barely any money left. The system was terrifying for rights holders because it created high overhead costs while generating uncertain returns. What were they to do when things went wrong? Each individual group had relatively few assets, so it was difficult for them to get loans for updated equipment or needed repairs.

The leaders in the salmon fisheries community saw a possible solution to what they saw as wasteful inefficiencies; if they convinced all of the salmon rights holders to join forces to create a single organization, they could eliminate redundancies and dramatically reduce their overhead costs. They could easily harvest the same number of nets with only a few collectively owned boats. And if the rights holders banded together, they would have more than enough people to haul in all of the nets themselves, saving them the cost of hiring and housing seasonal laborers. But convincing rights holders with long-standing rivalries and prideful independence to work together in new ways was not an easy task. While the fishers could see the benefits of collective organization, those with the best nets of the bunch worried that they might lose out if they joined with others. However, in 1994, after much deliberation, Kitahama's approximately 160 salmon rights holders voted to try out a radically different organizational structure, one that combined all the set-net groups into a single entity with one business office and jointly owned gear. Per Hokkaido law, each of the rights holders would continue to have stakes in their original net(s), but they would effectively sign over the rights to manage and profit from those nets to the newly formed salmon set-net cooperative. Instead of controlling their own nets, the rights holders essentially deputized a board of directors to manage their net in concert with all of the salmon nets in Kitahama.

Convincing people to take on new roles and to voluntarily give up direct control over their "own" nets demanded that the set-net leaders credibly conjure the new riches that such acts would generate. At the same time, they had to console some fishers who were saddened by the loss of their boats, which were sold off as the new board pared the number of vessels. They also had to assuage the egos of some rights holders who were initially reluctant to do the dirtiest work that they once assigned to the migrant workers. But perhaps the most challenging was that Motozumi-san and other leaders had to reassure the rights holders that they would not be cheated of their rightful shares of the pie—while also redefining what counted as rightful. Because certain nets had historically greater average harvests than others, the owners of those nets wanted bigger portions of the collective earnings than others. So too did the owners of nets with fewer rights holders and thus greater per-person earnings. But the rising co-op leaders had other ideas. They wanted to value *work* instead of historical privilege. To reduce overhead costs, they would have rights' holders staff the office, man the boats, unload the catch, and chase away the birds until buyers came to haul them away. Those who took on more tasks would get more money. The co-op leaders, especially Motozumi-san, saw systems that distributed wealth based on family ties and good fortune as fundamentally backward and those that rewarded work as more modern. While Motozumi-san was not a Marxist, he was a member of the Russian Club—a local study group that focused on Russian literature, culture, and politics—and appeared to draw worldly inspiration from thinking across different forms of economy.

MODERN BUSINESS

The question of what constituted modern business practices was a true question for Motozumi-san and his colleagues, one they contemplated through comparisons. They were inspired by American-style corporate governance structures, streamlining, and rationalization. After studying practices common in US businesses, they began charting which nets were the most productive and how much fuel it took to harvest fish from each trap and estimating the most profitable patterns for checking the nets. As they sought to make fishing a science rather than an art, data became king. Numbers about weather, water temperatures, fish population size, and boat usage were recorded on clipboards, displayed on dry-erase boards, and entered into computers. If a certain practice did not make sense according to available data, they changed the practice. In several cases, they stopped fishing nets that they found to be inadequately productive. When their number crunching revealed that they were spending too much of their gross income on buying ice to chill their fish, they built their own large-scale ice machine so they could eliminate the ice-maker middleman. In addition, based on the data they collected, they upgraded their boats, switching to vessels with higher-capacity fish holds and better fuel efficiency that had lower operating and maintenance costs.

Yet at the same time that they were attracted to American business practices, Motozumi-san and his cohort were also drawn to discourses about perils of the economic inequalities produced by unfettered US capitalism. They compared the poverty rate and lack of universal health care in the United States to the somewhat better safety nets of Japan and Europe. Under unrestrained capitalism, the fishers insisted, you end up with too much inequality. "Haven't you seen Michael Moore's movies?" they asked me, referring to documentaries by a popular director that focus on the injustices of American systems. But while they were wary of American capitalism, the fishers also felt that Soviet communism offered no ready-made solutions. "With 'pure communism' people get lazy," one fisher told me. "They don't work hard." Through reading, watching, and comparatively reflecting on traveling models in trade publications, business journals, and popular literatures, fishers sought to piece together bits of various systems to create what they saw as the right kind of inequality—a minor amount that motivated people to work hard without creating too many disparities.

Ultimately, they designed a system of fractional shares to divvy up the profits. For example, the top earner, the board president who bears ultimate responsibility for the co-op, receives a full share of "1," while the vice presidents might earn 0.92, or 92 percent of the largest share, and a hardworking man who volunteered to work on a boat might receive a 0.85 share, or 85 percent of the salary of the president. When Motozumi-san and the other co-op leaders talked it over, they decided that the "ideal inequality" would be for the average dedicated co-op member to earn about 0.80, or 80 percent of the top share. Such differences would reward people for taking on the risks and burdens of leadership without creating hard feelings.[11]

While this merit-based system proved more egalitarian, it lacked transparency. The spread of the fractional shares was public information. Every year, the set-net group gave each member a list of the distribution—twenty-five people at 0.83, thirty people at 0.80, and so on so that they could confirm that the general schema seemed equitable. Officially, members did not know which share others received; there were rumors, of course, but most people kept their share information secret. The actual allocation of shares was a cryptic process in which the set-net group's board members, in a closed meeting, privately decided each member's share. Once, I asked Motozumi-san if I could see a copy of the set-net groups by-laws, assuming that they had written rules and policies for determining who gets how large of a share and for determining who can inherit rights. "We don't have any," he replied. Initially I thought this statement was just a tactic to avoid sharing them with me, but I soon learned that he seemed to be telling the truth. Motozumi-san explained that the co-op board members had a shared sense of what was right and that they preferred not to be tied to any written rules. They needed flexibility, he said. The general principles for deciding shares were clear, he said: rights holders who do more demanding work receive larger shares, so that people who work on boats receive more than people who work on the docks. Within a particular category of work, effort and attitude count. For example, a boat worker with a reputation for being a hard worker would get a larger share than someone who chronically shows up late. They are punitive toward healthy but seemingly lazy young men who choose the easier dock work over joining a boat crew, but they are compassionate toward widows and people with physical ailments. During the months I spent at the co-op, I heard some minor grumbling about shares, but no serious dissatisfaction or dissent. Overall, the fishers were convinced that despite some small imperfections, they had come up with one of the most just and logical group structures possible for their circumstances.

IMPROVING PRICES

Once they were more efficiently organized, the fishers also began to seek better prices for their fish. Instead of accepting the abysmal rates for their salmon in Japanese domestic markets after the arrival of Chilean fish, the Kitahama fishers began searching for new buyers overseas who might pay more for their fish. Their salmon had been caught up in global market changes that had driven down their per-kilo value; in response, the fishers sought to make their salmon more worldly in order to thrive within these new economic conditions. One problem that they faced was that high

Japanese labor costs made the export of fully processed Hokkaido salmon to Europe or the United States virtually impossible. Japanese companies simply could not produce the frozen salmon fillets that consumers had come to prefer at a cost comparable to those of farmed fish. The Kitahama fishers thus sought out Chinese fish processors who were pioneering new supply chains, linking up with Chinese companies that bought low-priced, lower-grade, wild-caught salmon from around the North Pacific in a minimally processed state. These companies then cut the fish into single-portion sizes, deboned and repackaged them, and sent them off to European and American markets. China's lower labor costs and the convenient, ready-to-eat portions that the factories produced made otherwise lower-value salmon into a globally competitive product. Although the Chinese factories would not pay top dollar for Kitahama's salmon, they outbid the Japanese fertilizer plants and raised the price of Hokkaido salmon just enough that, when coupled with cost-cutting cooperative measures, they allowed the Kitahama fishers to reliably generate profits.

MAKING MODERN SELVES

These changes seem to have paid off for the Kitahama fishers. When I arrived in Kitahama in 2009, I encountered signs of salmon wealth. Gleaming stainless-steel boats decked out with the latest sonar lined a newly built concrete harbor, while the cars in the parking lot in front of the fishing co-op included several Audis, a couple of BMWs, and even a Mercedes-Benz. While this new money certainly allowed the fishers to cultivate the personal habits that marked them as part of a transnational cultured class, the fishers were not merely in love with their money. They also clearly loved the aesthetics and performance of being *kindaiteki*, or modern, as such. They loved their organizational systems and regularized patterns for rapidly and accurately sorting fish by grade and sex as they unloaded them from the boats. They loved that they had designed and ordered wonderfully efficient welded metal sorting tables and trained everyone to carry out their specific sorting job with an assembly-line mentality. They loved that everything on the dock had its place and that gear was always cleaned, stacked, and properly put away. And they loved that they had turned their fathers' cottage industry into an international export business.

It was Ohno-san who most clearly explained to me that it was the Kitahama fisherpeoples' worldly experiences and comparative thinking that had allowed them to achieve such success. Ohno-san sat on the floor of his living room cradling his pet Chihuahua in a failed attempt to prevent her

Hokkaido fishers remove salmon from a set-net's holding chamber. Photo by author.

from barking incessantly while we talked. Despite the inconvenience of the yapping dog, Ohno-san, a fifth-generation Kitahama fisherperson and the descendent of one of the town's Meiji era pioneers, very much wanted to talk. After attending an elite Jesuit boarding school, Ohno-san had majored in sociology at Hokkaido University and had written a bachelor's thesis on the social history of Kitahama's salmon fishing industry, a copy of which he eagerly loaned to me. Ohno-san's favorite word for describing Kitahama's salmon fishing practices was "evolved" (*shinkashita*), and the word appeared in virtually all our conversations. According to Ohno-san, Kitahama has the most advanced maritime technology of any Hokkaido fisheries group. But what really makes them most evolved is their way of thinking—they have overcome tradition (*dento*). "Other [fishing co-ops] just keep doing it one way because that's how they've always done it. We kept thinking that there must be a better way," Ohno-san explains. In contrast to other fishing groups, the Kitahama fishers, he says, are "able to see the world in different ways."

What Ohno-san calls being "evolved," other Kitahama fishers referred to as "modern" (*kindaiteki*). But regardless of the word they used, nearly everyone had the same explanation of what made them who they were: their ability to see the world from multiple perspectives and to compare across them. With such abilities, they are able to reinvent their relationships with each other and to market forces in ways that those without comparative thinking cannot. According to both Ohno-san and Motozumi-san, physically changing places had been essential in allowing them to think in worldly ways. "Living in Tokyo changed how I thought about everything—truly everything," Motozumi-san once told me. It gave them new models of being and new grounds for comparative thinking. When they returned home to Kitahama, their actions emerged out of their comparisons—between the practices of fishing co-ops and metropolitan corporations, between their lives in Kitahama and visions of who they might have become in Tokyo, and between their initially poor financial situation in Kitahama and their understandings of the resources they would need to cultivate the lives they desired. It was their comparisons between Kitahama and cosmopolitan elsewheres that motivated them to remake their fishery—and to do so in a way that made the fish matter as little as possible to them.

COMMODIFIED RELATIONS TO FISH

By the time I arrived in Kitahama, most of the fathers of people like Motozumi-san and Ohno-san had passed away or were in poor health. But their sons told me of the older generation's affection for the fish. One of the group leaders gave me a book of haiku that his father had written, which was filled with awe for the region's salmon and the sensuousness of fishing. In past generations, many Kitahama fishers told me, fishermen had an embodied, affective connection to their fish. However, they stressed that they did not yearn for such feelings or attunements. They did not want salmon to be lively parts of their lives, the stuff of their dreams and poems. Even if they could not completely distance themselves from Kitahama and its fishing industry, they wanted to do the best they could to separate themselves from the fish.

On one hand, the Kitahama fishers knew that their particular form of modern selfhood was completely dependent on salmon. Without the fish, their wealth and cosmopolitan lifestyles would be impossible. On the other hand, they strove to be businesspeople who dealt in data, not fishers who dwelled in the materialities of slime and flesh. But there was no getting rid of the actual fish. They were in their nets, on their boats, and on their docks. The Kitahama fishers' solution to this paradox was to the kill salmon as quickly as possible—not literally, but affectively. Even though the salmon were still physically alive when they hit the boat deck, as fish, they were already dead to the fishers. "I don't even really see them as fish," one fisher told me. "I only see them as money."

Like most Japanese people, the current generation of Kitahama fishers liked to eat salmon—grilled for breakfast, buried inside a rice ball for lunch, sliced into sashimi for dinner. But the fishers were not particularly fond of salmon as creatures. They rarely admired their fish or mused of the wondrousness of the salmon life cycle. When they unloaded the boats, each fisher had a designated task: operating the winch, opening the net chain to deposit the fish into the unloading area, moving the fish into the sorting areas with plastic snow shovels, or sorting the fish by species, sex, and grade. They wanted to get the job done as fast as possible so they could get on with their lives separately from the fish. If they kept on track, they could often be done with work by noon, leaving the afternoons free to play golf.

Before the formation of the unified Kitahama co-op, salmon were less easily contained, and the fish routinely spilled over into other parts of the fishers' lives. When they ran their own small set-net operations and were financially crunched, the fishers' needed to enroll the whole family in salmon-related endeavors. They needed wives and sometimes children to help unload the boat, manage the books, market their fish, and hand make products like salmon jerky that they could sell directly to tourists for a bit of extra cash. As part of their initial attempts to raise salmon prices, the fishers spent much of their free time organizing seafood promotional events such as an annual salmon festival complete with a "salmon derby" where people would bet on which fish would swim across a tank the fastest. After the formation of the unified co-op, the fishers stopped all such activities without an iota of nostalgia. "Traditional" fishermen had to do such things to stay financially afloat; "modern" fishing industry professionals did not. For the Kitahama fishers, salmon-centric lives were signs of failure. You hold salmon festivals and wax romantically about your connection with fish when your fishing business isn't going well, they told me. A well-run salmon group made enough money that its members did not have to spend their time on marketing gimmicks. Because they had overcome traditional modes of thinking to build a business-like salmon group, the Kitahama fishers were freed from having to perform tradition.

When I arrived in Kitahama, I immediately noticed how little the fishers' wives had to do with salmon. In other parts of Hokkaido, fishermen's wives were active participants in salmon worlds. They made *toba* (dried salmon) to sell at local stores and markets and set up food booths at regional events where they made and sold homemade seafood dishes. Sometimes, they also taught cooking classes, operated their own restaurants, or even ran direct-sales seafood stores. But in Kitahama, only a handful of women participated in the fishing cooperative's women's division, the entity through which fishermen's wives typically organize. I initially misread the Kitahama women's absence from fishing as a sign of potential oppression, and I asked a number of Kitahama wives if they were disappointed that they did not get to participate in the fishery. Was there something about Kitahama fishing culture that was preventing their participation? But as Motozumi-san's wife explained to me, I was missing the point. When fishing wives work, it is a sign of poverty, not empowerment. It is not that she has been excluded from the fishery; it is that she has the great privilege of not having to do so, because the Kitahama fishers are managing their fisheries well. As another fishing wife told me, Japanese fishermen typically struggle to find wives because women do not want to have to labor in the industry, but young men in Kitahama, who are able to keep their families separate from fishing, have much less trouble finding brides.

For the Kitahama fishers, making *kindaiteki* and *shinkashita* fisheries was a project of containment and transformation. They sought to contain the role salmon played in their lives by transforming them into abstract commodities as quickly as possible. As commodities, salmon could move, becoming cosmopolitan themselves. In doing so, they also generated the wealth that the Kitahama fishers used to surround themselves with the trappings—the commodities—of transnationally legible upper-class-ness. Motozumi-san's favorite story, which he told me several times, was about one of his trips to Europe. On a visit to Paris, he had arranged for a special tour of the central Paris fish market so that he could continue to expand his knowledge about the global seafood industry.[12] Much to his surprise, as he walked through the market's aisles, he stumbled upon a crate of Kitahama salmon. He had gone all the way to France only to encounter his own fish! I think Motozumi-san was especially fond of this anecdote because it demonstrated both his own cosmopolitanism and his success in turning his salmon into a global commodity. Through worldly thinking practices, he had successfully commodified salmon, freeing them to travel beyond local Japanese markets. In doing so, he had also separated himself from the parochialisms of fishing, instead building an identity as a businessman and creating the financial wealth and confidence that he needed to be able to travel to Europe. While the men of Motozumi-san's father's generation had known salmon primarily through bodily intimacies, Motozumi-san's worldly ways had enabled him to "know" Kitahama salmon from Paris.

Motozumi-san and the other Kitahama fishers liked the idea of salmon-as-commodities—as uniform units that they could convert to money and then to other goods—and they explicitly built the monetarization of salmon into their co-op practices. While salmon fishers in other parts of Hokkaido would commonly select a few of the most beautiful fish to simply take for their own tables and freezers, such practices were not allowed in Kitahama. If Kitahama fishers wanted some of their own salmon to take home, they had to buy them from the set-net group at the day's per-kilo auction price.[13] The moment they entered the set-nets, salmon were units of potential profit that belonged to the co-op. By requiring that everyone, including boat hands, buy their fish at the going auction rate, the Kitahama fishers intentionally closed the shortcut by which fish bound for their own tables had long bypassed commodification. Instead, they structured their practices so that salmon had to pass through a commodity-making apparatus—even for it to become their personal food.

The comparisons that the fishers made between their fathers and themselves, between traditional craftsmanship and modern business, and between salmon liveliness and commodity liveliness impelled them to rationalize salmon. Yet as much as the Kitahama fishers found pride in their objectification of fish, their alienation from fish was not complete. Their eyes still noticed differences among salmon, and they still felt something special toward the most perfect fish. One day, when I came home from the docks, Motozumi-san's wife, Mariko-san, was vacuum-packing salmon fillets at the kitchen table. The night before, I had heard them drawing up a list of *oseibo* (annual year-end gift) recipients, deciding who should get how much salmon and in what forms. Sending such gifts to family, close friends, and business partners is a common practice in Japan. But while most Japanese sent specialty food items purchased at a department store, the Motozumis sent their own salmon. During a lull in the morning action at the docks, Motozumi-san had brought home to Mariko-san some especially high-quality salmon that he had purchased from the set-net group. Most of the salmon who entered their nets were nearing their spawning areas and were thus beginning to sexually mature, losing some of their color, flesh texture, and fat reserves as they began to reconfigure their bodies for gonad development. However, they also captured a handful of sexually immature salmon with bright silvery skin and higher fat content (and thus more flavor). These special fish, called *keiji*, were said to be a one-in-a-thousand or even one-in-ten-thousand catch. They never appeared at regular supermarkets, but when I occasionally saw *keiji* for sale at high-end department stores or specialty markets, they were routinely being sold for the equivalent of about US$300 per fish. You could not tell for sure if a fish was a *keiji* until you cut it open and saw the absence of developed sex organs, but you could hazard a guess by looking at a fish's outward appearance. Some fisherpeople in other parts of Hokkaido would sort out the shiniest silvery fish that seemed likely to be *keiji* or other high-value immature fish such as *tokishirazu* (literally, "fish that don't know the time"), selling them individually at premium prices. But the Kitahama fishers did not seek out specialty markets, instead treating their fish as a mass product and sorting them into four simple categories, with one for female fish likely to contain roe and others for general quality grading. If there was a silvery immature fish, it might go right into a crate where it was buried among regular fish and sold at auction in bulk at the normal price per pound. But sometimes, such fish caught the eye of the fishers as they sorted the day's catch. They might take a moment's break from their work to grab that fish and stash it aside to buy at auction. Such fish were a bargain deal; they could purchase these salmon at auction prices that hovered around $3.50 per kilo, or about $25, depending on the size of the fish. These were the kinds of salmon that Motozumi-san handed off to Mariko-san, who then gutted, filleted, vacuum-packed, and froze the meat, while also preparing small plastic containers full of salmon roe (*ikura*) from female fish. The next day, she carefully packaged the frozen salmon and chilled roe in Styrofoam boxes and sent them through refrigerated mail. This process repeated itself for several days until each name was checked off the original list.

After the packages went out, the phone began to ring with expressions of gratitude from gift recipients. But one evening, when Motozumi-san answered, there was a different caller on the line: the refrigerated shipping company. They had some unfortunate news; the company had made an error, accidentally placing one of the carefully packed gift boxes into a regular mail truck rather than a chilled one. Because the product inside would no longer be safe to eat, the shipping company wanted to compensate Motozumi-san for the loss. Although I could only hear one side of the conversation, it was not difficult to imagine the other. "How much did you pay for the fish?" the shipping company representative must have asked. "I'm a fisherman [ryōshi], so I didn't buy the fish at normal price," Motozumi-san answered, identifying himself unusually as a fisherman rather than as a fishing industry professional. "Well, how much was it worth?" the company representative apparently replied. "That was the kind of fish you can't get your hands on, that you can't buy. It's irreplaceable," Motozumi-san said, emphatically. "It was a keiji. You can't calculate the value of a fish like that!" After more back and forth, the shipping company representative eventually offered an amount of compensation that I was not able to hear, and Motozumi-san, clearly still miffed, reluctantly accepted the settlement.

Most of the time, Motozumi-san was a fishing industry professional, and his fish were uniform commodities, known through spreadsheets and profit reports. Most of the time, he claimed that he did not like anything about fish. Once, when I asked him about what he liked best about working with salmon, he bluntly answered, "Nothing." He passionately claimed to be passionless about fish. But on occasion, Mariko-san would cut open a fish that was a little more silvery than the others and call out to her husband in a voice filled with wonder: "Keiji da yo [it's a keiji]." And in reply, even Motozumi-san would smile.

CHAPTER SIX

When Comparisons Encounter Concrete

Wild Salmon in Hokkaido

W HAT happens when one tries to make new comparisons in material worlds shaped by past ones?

Despite their efforts to be masters of comparison, the Kitahama fishermen have found themselves caught in comparative predicaments that they cannot easily solve. Not long after I arrived in Kitahama, I began to hear talk about the Marine Stewardship Council's eco-label. "It's what you need to sell at Walmart these days," one fisher told me. Established in 1997 through a partnership between the World Wildlife Fund and the seafood company Unilever, the Marine Stewardship Council (MSC) sets standards for sustainable fishing and seafood traceability. Products determined to meet MSC standards are allowed to use the organization's blue logo on their packaging. The mark is both a testament to a product's conformance to environmental ideals and a tool for increasing its value, as a growing number of consumers are willing to pay a premium price for eco-friendly foodstuffs. Many Kitahama cooperative members speculated that this eco-certification might raise the price of their salmon depressed by farmed fish and facilitate its travel into new international markets.

Believing their fisheries to be well managed and sustainable, Kitahama cooperative members thought that such certification would be relatively straightforward. Their hatchery was run with an eye to maintaining genetic diversity (e.g., by drawing sperm from multiple males rather than relying too heavily on a few), and they were not overharvesting their fish. In addition, the fishermen's cooperative had recently campaigned to keep a local lake clean, worked with upriver farmers to reduce pollution from agricultural runoff, and planted trees to protect local watersheds.

However, the Kitahama fishers soon realized that MSC operated with a definition of sustainability different from their own—one that stressed the management of *wild* fish. Under MSC policy, only salmon fisheries that include wild fish are eligible for certification:

> Given the MSC focus on the sustainability of global wild fish stocks, the concept of 'wildness' plays a central role in scoping enhanced fisheries. The fishery must incorporate some element of harvest of a wild population, and must be managed so that the natural productivity and genetic biodiversity of that population is not undermined with respect to any impacts on long term sustainability. . . . The intent is that management systems exist to control exploitation rates on wild stocks in order to allow for self-sustaining, locally adapted wild populations (i.e., adequate wild stock levels that can perpetuate themselves at harvestable levels on a continuing basis). (Marine Stewardship Council 2010, 5)

In short, for a fishery to be MSC eligible, it must have at least some wild fish, it must explicitly manage them, and it must work to rebuild any currently depleted wild stocks. These requirements have proved difficult for the Kitahama fishers. When the Hokkaido Federation of Fisheries Cooperatives initially approached MSC about certifying some of their salmon fisheries, they were told that they had no chance for certification until they had a wild salmon management plan. But Hokkaido fisheries professionals had never thought about managing wild salmon; indeed, the very idea of "managing" the wild seemed oxymoronic to them. Wasn't the very definition of the wild that which was not human-managed? They had invested in making a solid industrial production system with robust hatchery strategies. Wasn't that what good modern management was about? "We do have wild fish," one fishing industry professional told me, "but if we manage the hatchery fish well, the fisheries are fine. We don't 'manage' the wild fish. We don't have a policy about them." But as the Hokkaido salmon fishing community soon learned, such management approaches were not seen as acceptable in the eyes of either MSC certifiers or twenty-first-century Euro-American fisheries professionals. Proper management now required the careful counting, tracking, and support of a new management entity: the wild salmon.

Hokkaido's salmon communities have been caught in this abrupt shift in management logics. Throughout the late nineteenth century and most of the twentieth, good fisheries management was widely defined as that which increased harvests. From its inception, the very point of scientific management, be it of land or waters, was to maximize nature's bounty and harness it for productive use. By these measures, Hokkaido salmon fisheries managers excelled. By the late twentieth century, they had the world's most expansive salmon hatchery system, one that produced over 1.2 billion juvenile fish per year (Morita et al. 2006). For the Hokkaido's fishers and managers, such hatchery production has been at the core of their modernized salmon industry. But at the same time, this large-scale hatchery production system and the landscape changes with which it is intertwined also constrain the ability of participants in Hokkaido salmon worlds to enact an increasingly important part of contemporary environmental management: the protection of wild fish.

Since the mid-1990s, at the intersection of markets and science, the wild salmon—as a categorical entity—has taken the fisheries world by storm. When Chilean farmed salmon began to reconfigure global markets, they not only depressed prices but also created a new comparative categorical distinction between "farm-raised" and "wild" salmon. Alaska salmon fishermen, who had also been adversely affected by salmon price declines, jumped on a variety of reports about unsavory aspects of farm-raised salmon, especially their entanglements with chemicals and antibiotics used to manage fish disease. Drawing on Alaska's rugged image, salmon fishermen there began a campaign to distinguish their fish—some of which reproduced in streams and others of which came from hatcheries—as "wild" salmon. As Alaska fishermen began to promote their wild salmon as healthier and more environmentally friendly than farm-raised fish, fish designated as wild or wild-caught began to command higher prices in some markets in Europe and North America.[1]

Yet market changes alone did not bring "wild salmon" into being. New developments in salmon biology also shifted the aims and purposes of salmon management. Where management had previously focused on techniques for hatchery production and the maintenance of sustainable fishing yields, in the 1990s, a growing number of fisheries professionals asserted that the field must also aim to protect the genetic diversity and ecological functions of wild fish—in part by substantially reducing hatchery production. These changes were led by fisheries biologists in the continental United States and Canada, which were responding to their regions' late twentieth-century salmon crises, where stream-spawning salmon numbers were plummeting at the same time that new genetic technologies were revealing their significant distinctions. Increasingly, the point of salmon science was not merely to promote fisheries but also to protect fish as biological beings. In doing so, fisheries biologists began to transform their discipline from a science of production enhancement to one of wildlife conservation.[2] With such frameworks, Japanese fisheries, with their focus on hatchery production, suddenly seemed out of date. This new modernity demanded wild fish policies, not just hatchery strategies.

In the nineteenth century, Hokkaido officials struggled to make the island's landscapes legibly modern within definitions of modernity set by Euro-American powers, a process that included the development of salmon canneries and hatcheries. While the definition of modern salmon management had substantially shifted in the twenty-first century, the underlying structural pattern of modern comparison, in which North Americans or Europeans have continued to occupy the position of standard setters, has remained constant. While critical of such repetitions, some fisheries professionals in Hokkaido, especially younger fisheries biologists, welcome management approaches centered on wild fish conservation. But shifting gears and comparing anew has proven challenging. Hokkaido's rivers and fish have been fully entrained in projects of industrial production: the fish relocated to hatcheries and the rivers tamed to aid agriculture. As wild salmon management makes demands that rub up against existing institutions and landscape forms, Hokkaido fisheries managers struggle to confront the new calls: How can one craft a new form of modernity in a landscape fundamentally changed by previous modern projects? Their efforts to embrace wild salmon are constrained by literal concrete—the material remains of past landscape-making comparisons, which limit their ability to negotiate new ones.

THE INERTIA OF CONCRETE

Shinji Nakamura wants to return wild salmon to Hokkaido's Ishikari River, but it is hard work. A sixty-five-year-old retired high school teacher, Nakamura-san is the head of a local nature society that dreams of restoring wild salmon to the river's upper reaches, which flow through Asahikawa, the second largest city in Hokkaido, with a population of about 360,000. Nakamura-san was no newcomer to environmental advocacy; he had been protesting road-building, dam construction, and the overcutting of national forests for decades. "But I was tired of anti-this and anti-that activism. I wanted to make something, not just oppose things," Nakamura-san said. "That's what led me to salmon."

Since 1983, Nakamura-san has been trying to restore naturally spawning salmon to a place where they *should* be. In a nation where short, steep coastal rivers predominate, the Ishikari is one of the few that winds its way far inland, stretching 268 kilometers from its mouth on the Japanese seacoast to its headwaters in Daisetsuzan National Park. For millennia, salmon flourished in the watershed, and both archeological and historical records indicate that it may have had the largest salmon populations of any river in Japan.[3] Although some salmon climbed all the way to the river's tiny headwater streams, many spawned in its central stretches, located inside a large bowl-shaped basin called the Asahikawa *bonchi*. Because of the basin's unique shape, its waterways are hydrologically exceptional for incubating salmon eggs. Clean groundwater—percolated precipitation from the surrounding mountains—bubbles up through riverbeds, forming pockets of springwater that make perfect salmon spawning grounds, with stable year-round temperatures and high levels of oxygenation.

From the late nineteenth century onward, however, Asahikawa's salmon populations have faced more than their share of environmental insults. Traveling through the region by train, the view is one of seemingly endless irrigated paddies and fields. The river had to be diked and dammed to make such landscapes possible; marshes needed to be drained, violent spring floods controlled, and waters made available for agriculture. The remaking of this waterway has transformed Asahikawa into Hokkaido's largest rice producing region and promoted the expansion of other crops— but at great cost to its riparian worlds. The concretization of the river has been widespread; cement lines not only its sides but also its bottom, where it is used to prevent the river from cutting deeper into its bed and undermining its berms. Of course, the raw material for all that concrete had to come from somewhere, and that place, too, was often the river, which was mined for gravel. As Asahikawa industrialized, the river also became a convenient location for waste disposal. In the early twentieth century, the Japanese government established a wood-pulp processing plant upriver from one of the region's major salmon spawning grounds; soon, the river's water began to smell.[4] Despite this litany of damage, a handful of salmon managed to return to Asahikawa every year until 1963, when the construction of a downstream agricultural diversion dam without a fish ladder put a final end to their migrations.

The disappearance of Asahikawa's salmon was not a primary concern for twentieth-century fisheries management. To maintain fisheries, managers divorced salmon from their rivers; they collected the fish at river mouths, then moved them directly into hatcheries, minimizing their interactions with waterways given over to agriculture and industry.[5] From the perspective of many twentieth-century fisheries managers, the Asahikawa River's lack of salmon was not a crisis; hatcheries replaced its fish. However, not everyone agreed with this dominant management paradigm. Nakamura-san was one of the early dissenters. A lifelong lover of mountain climbing and wildflower identification, he was awed by ecological connections before such a position became popular. He was distraught by the ways hatcheries disconnected salmon from watersheds. "They [hatchery-produced fish] have no relation to rivers [*tsukiai ga nai*]," he says. "They are not real salmon [*honrai no sake de wa nai*]." "[Hatcheries] are just factories at the mouth of the river." Ecological connections, he claims, are what make salmon "real." One of his greatest dreams, he tells me, is for the bears in Hokkaido's Daisetsuzan National Park—one of the island's largest undeveloped tracts—to be able to dine on salmon once again as they did before the dam cut off the fish runs. The bears gesture toward a larger problem: that Hokkaido's watersheds are starving from a lack of salmon. Before hatcheries, post-spawning salmon carcasses would rot and be eaten by a variety of beings, including birds, mammals, and insects, their bodies fertilizing upland watersheds with the marine-derived nitrogen and phosphorous contained in their flesh. With hatcheries, these carcasses are primarily shunted into industrial processing chains rather than watershed nutrient cycles.

Nakamura-san is trying everything he can to make salmon once again part of ecological webs, but again and again, his efforts come up against the inertia of concrete. One Saturday, I joined Nakamura-san and other members of the nature society—mostly retirees, housewives, and families with young children—as they hauled rocks from a mid-river gravel bar to a side channel.[6] Their goal was to build a salmon spawning bed into which they might plant a few hundred eggs. But the most promising stretch of river they could find—a quiet stretch with some newly planted trees and some winding side-channels between relatively widely spaced dikes—was still decidedly unpromising. The diking had altered the area's groundwater flows and impaired the upwellings of springwater so essential for the survival of chum salmon eggs. "When you turn the sides and bottom of a river into concrete, you just don't have much left," Nakamura-san commented. Channelization had also increased the speed of the river's flows and thus scoured out much of the small-sized gravel that salmon need to spawn. Every year, Nakamura-san and his group nonetheless do the best they can. With shovels and bare hands, they build salmon spawning nests, selecting a place where at least a little fresh springwater seeps through the dikes. There, they dig a small channel, trying to shunt enough river flows over the eggs to keep them oxygenated but not so much as to wash them away. As a final step, they arrange rings of rocks to try to keep predacious fish such as rainbow trout (intro-duced from the United States) away from the eggs.[7]

The following week, they will return to the site with eggs from a local hatchery, burying the eggs into the nests they have prepared. The volunteers will also deliver some of the eggs to sixty local families, who will hatch them in aquariums in their homes, a practice that provides an opportunity for members of the public to learn more about the fish. Every year, at a com-munity ceremony, the families release their young salmon into a tributary stream, hoping that someday they will return to repopulate the river. In early April 2010, I watched as the families gathered just upstream from a bridge, where they had staked a bright banner with the slogan *Asahikawa wo yasei no sake no furusato in shiyō* (Let's make Asahikawa a hometown for wild salmon). There, engineers had built steps into the concrete flood-protection structures, making it easy for the parents and children to descend to the riv-er's edge. Each family had cared for about a hundred eggs, closely tracking their development over several months as they first developed eyes, then turned into tiny alevins (early life phase where newly hatched fish are fed by a pouch of egg remnants on their underbellies), then into hungry young fish. At the edge of the river, the families gently poured their fish from plas-tic buckets into the river while shouting *itterashai* (go and come back) to the young salmon, the same words one would use to send off a family mem-ber for a day at work or school. The chances are slim, however, that any of their fish will return. Even in the best of conditions, only about 2–5 percent of similarly sized salmon survive to reproductive maturity, and for these salmon, the odds are still worse. Among other challenges, the dam that killed off the region's salmon in the 1960s remains. Although it was retrofitted with a fish ladder in 2000, the structure was poorly designed and does not func-tion well. The entrance is too narrow and the water flows too quickly, form-ing small eddies and whirlpools that make it extremely tough for fish to ascend the river. "It's not enough, but it's all we can do," one volunteer says about the society's fish rearing and nest-building activities.

In 2003, Nakamura-san received the phone call for which he had been praying for two decades: a fisherman had found the decomposing bodies of two spawned out salmon along the bank of a nearby Ishikari tributary. Nakamura-san had those first two fish preserved in alcohol and put on dis-play at a local museum as a symbol of hope. If two fish could make it to Asa-hikawa, others could, too. It is a long way from two fish to a self-sustaining natural population, but it made him believe it was possible to restore some fish to the river. As they struggled to make their local waters more hospitable to salmon, Nakamura-san and the nature society did not invoke the kinds of cosmopolitan comparisons common to most other Hokkaido salmon contexts. Instead, they focused on *temporal* comparisons across pasts, presents, and futures as they cultivated a new sensibility: that watersheds where naturally spawning salmon have been extirpated are missing something that should be there.

At the same time, their work—undertaken in comparative idioms of past/ present and ecological connectivity/broken-ness—is underpinned by broader changes in transnational salmon worlds. During the years I spent in Hokkaido, Nakamura-san was also cautiously optimistic about his organization's new collaboration with a government-funded fisheries research institute. For years, Nakamura-san had prodded various agencies to provide him with larger numbers of hatchery salmon eggs to plant in the rivers around Asahikawa. Initially, no one was willing to consider doing so. Hatchery eggs were seen as a valuable resource, as the key raw material for their production; it made no sense to managers to waste eggs on a seemingly frivolous project that was unlikely to produce many harvestable fish. Yet in the midst of the changes in salmon management entangled with both the Chilean salmon-driven market shifts and the growing traction of comparisons between salmon conservation in Japan and North America, there is a new willingness to do so. As wild salmon promotion has become an increasingly important part of twenty-first-century transnational fisheries worlds, Japanese government agencies and research institutes that would have previously nixed the idea are willing to take up the restoration-oriented endeavors that have become a new comparative norm. These agencies are now more open to promoting stream-spawning salmon in the name of genetic diversity, and thus in collaborating with initiatives like Nakamura-san's volunteer-run project.

In 2009, the salmon research institute agreed to release young salmon into a tributary tied to Nakamura-san's project for three consecutive spring seasons. These releases were essential for Nakamura-san's group as they allowed for an infusion of fish on an otherwise impossible scale. The nature society can rear only a few thousand salmon through their network of household tanks and handmade spawning beds; in contrast, the government institute was able to release 250,000 juvenile salmon each year from 2009 to 2011, dramatically increasing the odds that enough fish would survive to reestablish meaningful spawning populations in the waters around Asahikawa. By autumn 2012, that indeed seemed to be the case; a friend in Sapporo phoned me in the US to tell me that he had just seen Nakamura-san celebrating on a Japanese Broadcasting Corporation program. Scores of salmon had returned to Asahikawa and were starting to dig spawning beds of their own.

Yet while large-scale releases created hopeful pulses of fish, their success remains uncertain. As the research institute's hatchery fish returned, trained observers documented more than 1,617 spawning nests in the area in 2012. But sadly, those nests did not lead to an equally robust second generation. By 2020, the number of returnees had fallen, and observers noted only two hundred nests. Even with more public and management interest in wild salmon, it seems that the ability of many of the rivers to sustain them remains limited. How can salmon eggs thrive in a river lined with concrete, where the waters no longer bubble? How can juvenile fish find anything to eat when dikes have replaced food-rich riparian borderlands? For new salmon populations to stick in the river, it might require hatchery supplementation for years, if not decades. And for self-reproducing salmon populations to survive, it could require the removal of the dam and the restoration of large areas of the river—major relandscaping projects for which there is not yet adequate political will. Asahikawa remains a landscape oriented toward mass production of grains, not fish. What kind of salmon populations are possible in rivers so tightly bound up with projects of industrial modernity?

THE CONSTRAINTS OF HATCHERY SUCCESS

Regardless of its limited success, the Asahikawa project exemplifies a mounting willingness in Hokkaido to experiment with wild salmon as a management category. Until the 2000s, there was no commonly used term for "wild salmon" in Japanese. Now, there are several, including *wairudo sāmon* and *yaseigyo*. The side-by-side use of these terms is emblematic of Japanese salmon management. The first word is written in katakana, a script used to mark terms as "foreign" loanwords. The second is written in kanji characters, a script originally borrowed from China but now largely domesticated as Japanese. As these two terms indicate, wild salmon are sometimes marked as a conceptual foreign import, but at other times, they slip into the language as a more Japanese concept. Hokkaido fisheries managers describe having a complicated and sometimes fraught relationship with wild salmon as a category that they take seriously yet whose accompanying conceptual frames sometimes seem overdetermined by others. Fisheries scientists in Hokkaido have contributed to scientific research on wild spawning salmon populations, differences in run timings between hatchery and non-hatchery fish, and genetic diversity among hatchery and wild stocks, to name just a few. But Hokkaido fisheries managers still feel caught in the comparisons of North American fisheries experts. No matter how hard they try, they cannot perform wild salmon management in ways that measure up to the ideals that emanate from North American fish worlds, a by-product of the material legacies of earlier comparative practices.

• • •

During the past decade, "wild salmon" has become a category of transnational environmental action taken up by North American–based salmon-focused scientific organizations and NGOs, which aim to conserve fish biodiversity through the protection of naturally spawning fish. Hokkaido fisheries managers participate in meetings of such groups, but they often find themselves on the margins as they struggle to make their salmon worlds legible. As one Japanese scientist explained, "For people who only want to protect wild salmon, Hokkaido just isn't very interesting." Hokkaido has comparatively few wild salmon, in part because Japanese officials put so much effort into modernizing and rationalizing them in the spirit of one kind of comparative modernity. Yet through the contingencies of evolutionary histories, they also had salmon stocks that were more amenable to hatchery cultivation (at least for a while) than were many in North America.

The material affordances of Hokkaido's salmon proved exceptionally well suited for industrial production in the late twentieth and early twenty-first centuries. Chum salmon, which currently account for more than 90 percent of salmon harvested in Japan, proved comparatively easy to cultivate once one had a basic understanding of fish nutrition and migration. After they hatch, young chum spend only a few weeks in freshwater before migrating to the ocean, in stark contrast to the extended freshwater residence times for sockeye (ranging from one to four years), coho (typically about a year), and Chinook salmon (ranging from three months to about a year), species that predominate in much of North America.[8] Such quick migration is a useful adaptation for life in Hokkaido's rivers, which are generally short, steep, and prone to high-velocity floods, as juvenile fish need to be ready to be swept downstream shortly after they emerge from their rocky nests. These high-speed rivers tend to have few marshy wetlands and provide relatively little food for fish, so it is also advantageous for salmon to quickly leave these barren natal streams to seek out the bounty of the sea. This short freshwater residence time makes Hokkaido chum a good fit for industrial hatchery production. Held in tanks for only a few weeks, there is less time for the fish to contract diseases or become dependent on human feeding. Such traits also make chum cheap to rear, as they have low feed costs and hatcheries can get by with hiring temporary staff to care for the fish during the brief time they are there. The fish are also so small that millions of them can be kept in relatively little space, lowering per-fish production costs.

Hokkaido's geology and geography offer still other advantages to fish cultivation. In early life stages, salmon growth is closely linked to water temperature. In Hokkaido, groundwater temperatures are warmer than the autumn and winter temperatures of many northern latitude rivers, so the water that hatcheries circulate through their pipes helps the fish to develop more quickly than in many other regions, giving them a developmental jump-start. Hokkaido salmon also have a migration pattern that is especially kind to young fish. Hokkaido salmon spend the first year of their lives feeding in the Okhotsk Sea, where the high nutrient inputs from the Amur River allow zooplankton to flourish, providing young fish with a bountiful buffet. Japanese chum feed in this watery nursery for a year before moving offshore in search of bigger prey. As Hokkaido hatcheries ramped up production, this gentle entry into the world proved ideal for hatchery fish to transition from industrial facility to open ocean, ultimately increasing fish survival and enabling Hokkaido hatcheries to become profitable ventures. Hatcheries, however, were not equally successful for all parts of Hokkaido; they worked best along the Okhotsk seacoast, where the number of returning chum jumped from an average of 1,369,000 annually during the years of 1959–70 to 16,998,000 annually during 2008–12—about 12.4 times more fish (Kitada 2014). Researchers and managers widely attribute this success to the regional hatchery production association's innovations in hatchery technique, including better egg handling and juvenile release times more closely matched to ocean conditions (mimicking natural out-migration patterns) (Kitada 2014).

While the concretization of rivers is one material force that wild salmon management must confront in Japan, this success of hatcheries is another. Today, the comparative practices that birthed and shaped the Hokkaido salmon hatchery system have been transfigured through their daily production practices into the bodies and population structures of the island's salmon. Since 1980, Hokkaido's approximately 150 private and ten national hatcheries have consistently released about one billion juvenile salmon annually into about 140 rivers and from eighty temporary net-pen rearing sites, an act that in its repetition has substantially modified the region's fish (Hokkaido National Fisheries Research Institute 2020; Kitada 2014).[9] Because

A Hokkaido salmon hatchery's outdoor rearing ponds. Incubators and smaller fish-hatching ponds are located inside the building at the back of the picture. Photo by author.

they usually return to the river of their birth, salmon develop distinct populations in each of the waterways where they live, with genetic, morphological, and behavioral differences attuned to that river's specific conditions. Hatcheries, however, have long swapped fertilized eggs among them, transplanting fish from one part of Hokkaido to another and scrambling the links between fish and their rivers; such processes of homogenization have left Hokkaido's salmon with weak remnants of what was once likely substantial variation among multiple regional populations (Sato et al. 2014). At the same time, by disproportionately releasing salmon juveniles in certain rivers over others, they have also redistributed the salmon that remain. Areas without hatcheries often have very small numbers of salmon, while those with hatcheries have a superabundance of fish—albeit fish who are mostly either caught in commercial fisheries or trapped for hatchery use rather than dying in streams, feeding other animals, and thus participating in ecological relations. Furthermore, because hatcheries have repeatedly propagated the earliest returning fish as they have rushed to fill their incubators as soon as fish runs begin, their labors have consistently selected for early returning fish, thereby moving the peak of the salmon spawning season earlier in the year (Sahashi 2021). These are merely a few examples of the changes that Hokkaido hatcheries have produced in the evolutionary trajectories of these fish.[10] Over their nearly 150 years of operation, they have inscribed the quotidian practices of comparative industrialization and modernization into flesh, genes, and relational worlds with durable and intransigent material effects not unlike those of dams and hardened river embankments.

In contrast, in much of North America, hatcheries simply did not work as well. This was particularly true in places like the Columbia River basin.[11] In chapter 2 we saw how the Columbia River served as a key comparative site in the making of Hokkaido's salmon canning industry in the nineteenth century. In the late twentieth and early twenty-first centuries, it has played a new role in the emergence of the forms of wild salmon management—which call for reductions in hatchery salmon releases—against which Hokkaido fisheries managers worry that their fish and landscapes cannot compare well. The development of wild salmon as a management category (not simply as a market category vis-à-vis farmed fish) is itself partially attributable to the failure of Columbia River hatcheries. Like their counterparts in Hokkaido, Columbia River officials constructed a number of salmon hatcheries in the late nineteenth and early twentieth centuries, touting them as a technology that would augment the region's fish runs. But while hatchery managers released hordes of juvenile fish every spring, the return rates of those fish remained dismally low, as evidenced by ongoing total salmon declines. In the 1960s, US federal and state agencies invested in hatchery research, as the Japanese government had done, but found less success. Why did hatcheries in Japan become so productive while those in North America did not?

The specificities of the fish, it turns out, mattered tremendously to hatchery success. The lengthy freshwater residence times for Chinook, coho, and sockeye, the species that composed the majority of Columbia River stocks, make them less ideal for ranching, in contrast to chum, which migrate to sea soon after hatching. When cultivated in a hatchery setting, these species require extensive feeding and care prior to their ocean migrations, and this extended freshwater phase magnifies the challenges and costs of their rearing. Diseases can more easily sweep through tanks and thus must be more carefully managed and treated. Fish behavior also changes as young salmon became accustomed to lives in which they swim in concrete raceways and food falls from above on a regular schedule. When they are released from hatcheries, some salmon seem to confuse the movements of predatory birds with the hands of the hatchery workers who would sprinkle food pellets into their tanks; rising to the surface of the water expecting to find food, they themselves are often devoured. Holding fish in hatcheries for months also proved expensive. The cost of the feed was one factor, but so was staff time. In contrast to Hokkaido, where young salmon demanded only a few weeks of care each year, Columbia River hatcheries need to pay year-round staff to do a range of labor-intensive tasks, such as tank cleaning, that must be repeated again and again during the fishes' long hatchery period. Moreover, once they are released from hatcheries, Chinook and coho rely much more heavily on riverine and estuarine habitats than do chum, who head directly to the ocean. Yet those essential habitats have been hammered by flood prevention and drainage practices throughout the past century, with an approximately 74 percent loss of estuary wetlands—prime feeding areas for these juvenile salmon before entering the ocean—thus leading to less robust fish and higher mortality rates (Brophy et al. 2019).[12]

Due to the coupled effects of species differences, hatcheries yielded much less impressive results in Oregon, Washington, and California than they did in Hokkaido.[13] By the 1990s, the profound failure of these hatcheries became clear in the Columbia River, as salmon populations dipped to such catastrophically low levels that several subpopulations of the basin's salmon were placed on the national endangered species list, and commercial fishing was so severely curtailed that fishermen turned to federal disaster relief funds to pay their bills. As fish numbers plummeted, distrust of and distaste for hatcheries began to rise (Cone and Ridlington 2000). In the Columbia basin as in Hokkaido, hatcheries were integral to the dream of "salmon without rivers," of being able to sustain a flourishing fishery while freeing rivers for other uses (Lichatowich 1999). In the Columbia, however, fishermen and fishing-dependent towns ended up feeling duped. They gave over their rivers to the schemes of government planners, but they did not get the bountiful salmon runs they had been promised. As salmon populations bottomed out in the 1990s, scientists and the general public became alarmed by declines in wild fish populations. The Redfish Lake sockeye, a unique subspecies that migrates more than nine hundred miles inland to Idaho and turns bright red when it spawns, became the "poster fish" for such losses. In 1992, only a single male fish returned to Redfish Lake, drawing mass media attention. Sometimes compared to the last surviving passenger pigeon, the fish was given the name Lonesome Larry, sparking political appeals for new forms of fish conservation (White 1995, 104). That same year, the Redfish Lake sockeye, along with a number of other salmon populations, were listed as endangered species under US federal law, an act that brought salmon into legally mandated management schemes as wild animals. Because endangered species laws are designed around logics of nature protection not commodity production, fisheries biologists and legal experts decided that only non-hatchery fish should be eligible for its benefits including wide-ranging safeguards against habitat damage and excessive harvest.[14] Fisheries managers suddenly found that they needed to reform all of their management strategies, aligning them with endangered species goals of sustaining wild fish, a new entity that they sought to better research and define.

An increasing amount of such research indicated that hatchery salmon themselves might be exacerbating declines in wild salmon numbers. Instead of returning to the facility of their birth, some hatchery salmon inevitably strayed and spawned in creeks, where they mated with wild salmon, diluting their genetic diversity with that altered by decades of hatchery production. Scientists began to fear, with good cause, that hatchery fish strays would endanger salmon diversity and well-being. Research was beginning to show that salmon that lacked local adaptations to their specific watershed showed a significant reduction in fitness, meaning that the descendants of mixed hatchery and wild salmon parentage produced fewer offspring than did fish of exclusively wild parentage (Araki et al. 2008). Through this and other mechanisms, hatchery fish were likely edging naturally spawning salmon populations closer to extinction. In response, new approaches to wild salmon management have included attempts to spatially segregate wild and hatchery fish so that they do not breed with each other, alongside calls for reductions in hatchery salmon production in some contexts. For example, in 2013, wild-fish advocates successfully sued in US federal court to reduce Chinook salmon hatchery releases by a third in a Columbia River tributary, and more generally, in 2017, government fisheries scientists recommended that releases of hatchery fall Chinook in the Columbia drop from about eighteen million per year to about fourteen million per year (Learn 2013; Milstein 2017). In Oregon's smaller coastal rivers, which are closer in size to many Hokkaido waterways, some hatcheries have been shuttered entirely. In the 1990s, the Oregon Department of Fish and Wildlife reduced annual releases of coho salmon in such rivers from between four and five million to fewer than a million, while in 2007, it discontinued additional hatchery programs. For a couple of years after closing hatcheries, fisheries officials often trap and kill hatchery fish when they return to the river—rather than allowing them to spawn—to prevent hatchery fish genes from adversely affecting wild fish populations, even when that meant sharply reducing the total number of spawning salmon. In one controversial case, Oregon officials decided to kill and remove about 6,600 hatchery-origin salmon from a creek after a hatchery closure instead of allowing them to breed on their own, to protect the genetic potentialities of the thirty-two wild fish that also sought to spawn in the creek (Foster 2002).[15] In places where hatchery programs continue, managers sometimes attempt to spatially separate hatchery and wild fish to foster the harvest of hatchery fish while minimizing that of the wild fish. In the lower Columbia River, for example, hatchery fish are specifically reared and released in side-channel areas to entice them to return to those sites, where commercial gillnet fishing is allowed, while reducing the capture of wild fish who most frequently swim up the river's main channel.[16]

Within such contexts, US Pacific Northwest biologists, fishermen, and members of the general public began to harbor serious wariness, if not passionate dislike, toward hatcheries and hatchery salmon, which became symbolic of a broad overreliance on technological fixes and a neglect of ecological integrity. An essay by a US Northwest fisheries biologist illustrates the sentiments that became common in these salmon worlds:

> The real danger of hatcheries and other forms of artificial production is that they provide an excuse for habitat loss and poor fisheries management. If we believe in hatcheries, then we can allow the rivers to be dammed, silted, and destroyed. Just mitigate with a hatchery. Although it seems we should know better by now, the lessons haven't really sunk in. . . . Even more dangerous is the spread of the technological fix syndrome. If hatcheries don't work now, we will try some other form of hatchery technology. (Hilborn 1992, 7)

HATCHERY SALMON IN HOKKAIDO

In Hokkaido, however, a contrasting set of hatchery salmon policies and sensibilities around hatcheries developed in the late twentieth century in light of its industrialized salmon system, which functions relatively well. This moderate success has jointly material and affective consequences, as it is harder to be troubled by that which appears to work. Keiichi Yamada is among those who insist that hatcheries have saved Hokkaido salmon, not contributed to their declines. In his late seventies, Yamada-san is a retired hatchery researcher and director who held numerous salmon-related positions across Hokkaido during his thirty-five-year career. Unable to fully retire, he subsequently served as the director of a salmon museum, published a trade press book on salmon, and sat on the boards of at least four salmon-related nonprofit groups. If it hadn't been for hatcheries, Yamada-san tells me, Hokkaido salmon would have gone extinct. During World War II and the immediate postwar period, hatcheries were what protected salmon from poaching. People were hungry, Yamada-san says. They would have harvested every last salmon from the streams. But the fish were saved because the hatcheries collected them at the river mouths before poachers could get their hands on them and brought their gametes into protected locales, before selling their flesh onward.

In Hokkaido, hatcheries are also widely credited with making more sustainable salmon fisheries. Many people, including Yamada-san, describe the 1950s and 1960s high-seas salmon fishery as an example of an environmentally exploitative industry; although seen as "necessary" in the context of postwar food shortages, the factory ships are commonly described as a bad system that only harvested, without giving back. The transition from high-seas salmon fisheries to coastal harvest of hatchery stocks was widely celebrated as a switch "from fisheries that take to fisheries that make" (*toru gyogyō kara, tsukuru gyogyō e*), a line repeated again and again in cooperative association materials and by fishermen themselves. Hatcheries, which made the entities that fishermen caught, were at the heart of notions of responsible fisheries. For most Japanese salmon industry people I met, the financial structure of Hokkaido hatcheries were also key to their sustainability. In contrast to the Columbia River, where salmon hatcheries are primarily funded through federal allocations (with minor contributions from general state budgets and license fees), Hokkaido hatcheries are directly funded by the fishermen, who pay a set percentage of their gross income to their regional hatchery network each year. The system is seen as a stable cycle in which profits from the harvest of hatchery fish fund the production of the next generation of salmon. Overall, in the context of booming salmon population numbers and international pressure to abandon high-seas fishing, people in Hokkaido came to experience hatcheries and hatchery fish as good rather than as dangerous and inferior to wild salmon.

MATERIAL DIFFERENCES

The different histories in Hokkaido and the Columbia River have produced different material possibilities for wild salmon management. While it is important not to romanticize the state of the Columbia River, a watercourse I once heard described as "lobotomized," it nonetheless offers more affordances for enacting wild salmon than do the rivers of Hokkaido. On the Columbia, hatchery production was never as comprehensive as it was in Hokkaido. Weirs, for example, were not used to block salmon runs and collect fish near river mouths, in part because of the differences in the hydrology of the Columbia and its physical size; the river—1,243 miles long and draining an area about the size of France—is more than four miles wide at its mouth. Furthermore, while the Columbia and its Snake River tributary are dotted with eighteen large hydropower, transport, and irrigation dams, its tributaries have not been as extensively modified as those in Hokkaido. The streams of the Columbia River basin have been routinely silted by runoff from logging sites and stripped of their riparian vegetation, but they have not been subject to the mass construction projects that have turned Hokkaido's waterways into channels of concrete.

These material conditions have made it possible to implement elaborate forms of wild salmon management focused on the protection of existing wild fish stocks with at least some hope of success. As wild salmon have gained traction as an object of concern in the Columbia River basin, state and federal agencies have designed ambitious policies to try to save them—policies that alter water quality standards, fish harvest regulations, hatchery management policies, scientific research priorities, and hydroelectric dam operations. These new regulations have had profound effects on regional agricultural, logging, and ranching practices, requiring a reduction in irrigation water, forest buffers to shade streams, and fencing to keep cattle from trampling spawning grounds. Inspired by the material conditions of rivers and salmon, environmental managers and advocacy groups have used the legal frameworks of the US Endangered Species Act, which privileges species rights over economics, to compel such changes, even when they have severely curtailed industry. In the case of logging, efforts to protect wild salmon have reshaped policies and limited cutting to such an extent that I once heard a resource manager quip that Pacific Northwest logging regulations are "basically a salmon management plan." Since the 1990s, salmon have impacted a large fraction of decisions around land and water use in the Columbia basin; they determine who gets to fish, how much power gets generated, and who gets to develop property.

Since the early 2000s, the category of "wild salmon" has become ubiquitous across wide swaths of North Pacific salmon worlds. It has come to dominate international salmon scientific conferences and make a claim to a sizable portion of the global higher-end salmon market share. As talk of wild salmon has proliferated in North America, it has created ripples that salmon managers in Japan cannot ignore. As in the nineteenth century, international legibility and comparability still matter in Japan, and just as during the colonial settlement of Hokkaido, practices of land-use and environmental management continue to be key modes for enacting modernity. In the intervening century, the goalposts for "modern" human-nature relations have certainly shifted—frontier expansion is out and environmental conservation is in—but the desire to compare well remains. But how can one compare well when one's landscape and fish have been more severely altered and one's legal and policy frameworks have fewer affordances for protecting stream-spawning salmon?[17]

WILD SALMON IN JAPAN

Despite more than a century of intensive harvest, hatchery production, and river channelization, there are still some stream-spawning chum salmon in Hokkaido. Until around 2010, it was widely assumed that nearly all (greater than 95 percent) of Hokkaido chum salmon were of hatchery origin, and there was little optimism around the possibilities for substantial stream spawning (Kitada 2014; Miyakoshi et al. 2013).[18] Yet in 2008, island-wide surveys for chum documented their presence in 104 of 239 streams (Nagata et al. 2012), including sixty-five rivers that have never been stocked with hatchery fish (Nakagawa 2009). Revised estimates have indicated that while wild-spawning salmon were thought to make up less than 5 percent of total chum production, that number may instead be a little over 25 percent (Morita 2014).

Other research has also shown that despite decades of hatchery influence and significant homogenization, Hokkaido's fish populations indeed retain *some* genetic and life history diversity, especially those fish who spawn in rivers, often later in the season, and that may have been less incorporated into hatchery regimes (Beacham et al. 2008; Nagata et al. 2012; Sato and Morita 2019). But for the Kitahama fisherpeople, these discoveries did not seem enough for them to make internationally legible wild salmon in Japan without broader structural management changes. In 2014, after a series of private exchanges with the Marine Stewardship Council, the fishers decided to withdraw their application for its eco-label. The Hokkaido fishers had many doubts about the program, including the degree of financial benefit, especially after Alaskan salmon fisheries announced their withdrawal from the MSC program in 2012. In principle, groups such as the Kitahama fishing professionals supported the calls to support wild fish; they consistently embraced the restoration of their local rivers, orchestrating stream habitat initiatives that included dam removal, streamside tree planting, and erosion control.[19] From the early 2000s onward, they had started to collaborate with dairy farmers to keep cow manure out of waterways and with organic farmers work to reduce use of agricultural chemicals.

Like the Kitahama fishing professionals, participants across Hokkaido salmon fisheries rarely reject wildness or wild salmon conservation per se.

"As it's being said internationally, biodiversity [*seibutsu no taiyōsei*] and eco-system conservation [*seitaikei no iji*] are the point of view from which we are going forward," one former hatchery manager told me. In Japan, "those of us who are connected to salmon are thinking about them as wild animals [*yasei dōbutsu*]. We're looking for hatchery production that approximates wild [*yasei ni chikazukeru*]. We can't say that just because hatchery produc-tion has succeeded everything is fine." Yet this manager and others with whom I spoke are slightly wary of North American wild-based management regimes—including those embodied in MSC policies—that might displace hatchery production and threaten Japanese enactments of salmon-as-food. "The Japanese are fish-eating people," Yamada-san explains. "With our small land area, we can't really rely on grazing, so we need to ranch the sea." North American approaches to wild salmon management, which call for reductions in hatchery production to protect the genetic diversity of fish and reduce competition for food in ocean feeding grounds, worry them. While there are currently plenty of Chilean fish on store shelves, should Japan risk cutting back on its limited food resources? Although Japan has not suffered from famine for more than half a century, memories of postwar starvation remain alive, and food insecurity remains the stuff of everyday parlance, with fre-quent newspaper articles about low levels of domestic food production.

Yamada-san, the salmon education instructor we met in chapter 4, is among those who want to balance notions of salmon-as-wild-beings and salmon-as-food. In addition to his other activities, he organizes an annual Japanese-Canadian youth exchange program focused on fish. In odd years, Sapporo students travel to a town in British Columbia to learn about human and salmon worlds on the other side of the Pacific. In even years, the British Columbia students come to Sapporo, where in 2010, I tagged along with the visiting group of Canadian junior high students. In addition to general sight-seeing, their schedule included a visit to a salmon hatchery, a trip to a salmon museum, a fish dissection, and several conversations about nature and fish. Yamada-san wants the Canadian students to get a sense of how Japanese salmon-human worlds are different from their own. "Salmon equal nature in Canada," Yamada-san tells me. "They even cut down trees and leave them in rivers for salmon." Yamada-san wants the students to see why such wild-based management practices are so difficult in Japan's concrete rivers. He wants the Canadian kids to understand the predicament of Hokkaido salmon—that in the legacies of nineteenth-century modernization and ongo-ing limitations on food production, there are no quick fixes for Hokkaido's damaged rivers and displaced fish. If they can understand Japan's challenges, he says, perhaps the Canadian students will understand that the Japanese are not people who fail to understand the importance of nature conservation but instead are trying to do the best they can with their heavily modernized landscapes.

Japanese fisheries scientists are among those who feel marginalized by dogmatic approaches of many wild-centric North American fisheries experts. The scientists with whom I spoke are well respected in the trans-pacific fisheries science community, routinely attend conferences in North America, and publish their work in English-language journals. But they are irked at how North American salmon experts get to define what count as "best practices," while Japanese salmon professionals are criticized for failing to conform to the wild-centric values of North America. To them, such attitudes are ironic, even hypocritical. "*You* gave us hatcheries," one scientist reminded me. In the descriptions of several Hokkaido fisheries professionals with whom I spoke, the United States is not a "savior" of Japanese wild salmon but one of the chief causes of their decline. Nineteenth-century power relations forced Japan to emulate the West's flawed approaches to natural resources. One scientist showed me a chart that indicated that nineteenth-century Japanese started to build hatcheries in Hokkaido before salmon populations started to decline. According to the scientist, Japan built hatcheries because they needed to appear *kindaiteki* (modern) rather than because they needed to supplement salmon numbers; the Meiji government desperately did not want to be "behind" (*jidaiokure*). In an attempt to enact modernity through natural resource management, the Japanese abandoned effective but "passive" (*ukemi*) resource management practices, like protecting spawning grounds, in favor of more "active" (*sekkyokuteki*) modes of management, including Western-style hatcheries, even though they may have hastened declines in Hokkaido salmon populations by removing spawning from still healthy nineteenth-century rivers and putting their gametes into then-unproductive hatcheries. Before the West started meddling, the first scientist emphasized, the Japanese had their own salmon management systems based around the protection of salmon spawning rivers. This *tanegawa* (seed river) system strictly enforced fishing bans in certain rivers to ensure adequate salmon reproduction. Thus, North America, this scientist asserted, is not the only place that can stake claims to a history of wild salmon protection. Japan, he stresses, has its own genealogy of conservation in which human cultivation is not a constitutive outside.

As the same fisheries biologist explained to me, the sentiment popular in places like the Columbia River that "because stream-spawning salmon are good, hatchery fish are bad, well, that argument just doesn't make sense to us." As another scientist put it, protecting some wild spawning creates an important genetic reserve "so if anything goes wrong with the hatchery fish" there is some place to turn. "You don't need a lot [of naturally spawning fish] for that." "It's not that we don't like wild salmon," he continues, it's that too much of a focus on the wild just is not practical. According to the best available estimates, at its peak, pre-hatchery Hokkaido salmon fisheries produced only about three million fish per year, while in the 2000s, hatchery-based fisheries produced about sixty million fish (Kobayashi 2009, 13). Within Hokkaido, this temporal comparison—of recent past of scarcity to substantial abundance—comes into conflict with the wild salmon comparisons that move between Japan and the United States.

As another Hokkaido salmon scientist pointed out to me, although the wild salmon policies promulgated in places like the Columbia River claim to be working toward sustainability, they are radically unsustainable in practice. "There's an irony to it," he said. "So all of you people in Oregon and Washington, the salmon you eat, it's all coming from hatchery fish in Alaska. You only want to protect your own salmon as wild. To me this is really strange. You only care if your own place is wild. Other places it's whatever goes." While he overstates the proportion of hatchery salmon in Alaska, where around 30 percent of harvest fish are of hatchery origin, he nonetheless has a point; at the same time that fisheries professionals in Oregon, Washington, and California criticize Japanese fisheries managers for neglecting nature conservation in favor of production, the continental United States consumes a large volume of salmon from less than sustainable places, including Chile, where the United States has overtaken Japan as the largest destination for Chilean farmed fish.

Since around 2018, however, this confidence in Hokkaido hatchery salmon has been on the wane. While Hokkaido salmon returns ranged from about thirty-nine to forty-eight million fish between 2008 and 2011, those numbers had roughly halved by 2018–20, when only eighteen to twenty-three million returned, a level not seen since the early 1980s (Hokkaido Fisheries Research Institute 2020). These sharp declines have unsettled the fishing industry and tempered managers' celebrations of hatchery success. While the reasons for these low numbers are not fully understood, fisheries scientists think that the warming water temperatures wrought by global climate change are likely the primary cause. As the optimal water temperatures for salmon shift farther north, Hokkaido fish seem to be suffering, while more northerly populations in Russia see improved survival rates and record high harvests. Some scientists also fear that hatchery production has made the situation worse for Hokkaido's salmon, as the reduced diversity of hatchery fish gives the overall population fewer options for adapting to warmer oceans; a few have even posited that hatchery fish may be less physically fit than wild fish and thus less able to survive under the additional rigors produced by climate change.[20] Furthermore, data from one major Hokkaido river indicates that in years of low returns, hatchery fish tend to suffer proportionally more substantial declines than do wild fish, pointing to the likely ability of wild fish to cope with a wider range of conditions.[21] With these mounting worries, Hokkaido fishermen's cooperatives have ramped up their watershed conservation efforts, with their annual number of salmon-specific restoration initiatives leaping from thirty-three in 2008 to 122 in 2018 (Japan Ministry of Agriculture, Forestry and Fisheries 2020, 84). Yet larger policy shifts and governmental actions remain limited, especially when it comes to removing concrete from rivers. For example, despite the support of many Japanese fisheries biologists, it has taken nearly fifteen years to make substantial progress on the Rusha River, located within Hokkaido's Shiretoko UNESCO World Heritage Site. By 2019, the river's hatchery had been decommissioned, and two of three concrete dams on its Rusha River had been disassembled down to their river's waterline, but the removal of their remaining structures are expected to take several more years (DeNies 2019; Rand 2020). Although concrete structures in the park's other rivers have been notched to improve salmon migration, with subsequent increases in upriver fish spawning, they remain in place. The sluggishness of such projects points to a more general challenge: the structures of improved rivers and hatcheries are concretized in practices of governance, as well as in the concrete of the dams, dikes, and hatchery ponds themselves.

COMPARING WITH CONCRETE

Tensions manifest as Hokkaido's salmon management efforts are pulled into implicit and explicit comparisons with those of North America, sets of comparisons in which the distinctions between wild and hatchery fish have reconfigured the terms on which they are made. These comparisons crop up in a variety of fisheries management contexts, not only in the previously discussed attempt to secure Marine Stewardship Council certification but also in transnational ocean fora, NGO initiatives, and United Nations heritage-site plans, where various international groups pressure Hokkaido managers and government officials to remove dams, improve fish spawning habitat, and reduce hatchery production. To a certain degree, such pressures are not bad; there are substantial numbers of people in Hokkaido, including members of the Asahikawa River nature society and numerous fisheries scientists, who would like to see the Japanese government invest more assertively in river restoration efforts. These tensions also appear in international ocean management contexts, where North American scientists charge Hokkaido hatchery chum with adversely affecting North American fish who feed in the same ocean areas. They assert that the extra competition for limited quantities of prey generated by the large numbers of hatchery fish is reducing the survival and body sizes of Alaskan fish, which include a substantial number of wild salmon (Cunningham, Westley, and Adkison 2018; Ruggerone, Agler, and Nielsen 2012; Ruggerone and Irvine 2018).[22] Viewing these stocks as more valuable than their mass-produced Hokkaido counterparts, they call for reductions in Japanese hatchery production to improve conditions for wild fish. Again, such concerns also have their merits, as it does indeed appear that the total number of salmon is running up against the limits of the ocean to feed them all.

Yet in such contexts, Japanese salmon worlds are frequently depicted as behind the times, as needing to catch up to what North America–based fisheries professionals see as the proper approaches to wild salmon. Given the state of their rivers, their fishing industry built around hatchery fish, and the political and practical challenges of removing concrete, Hokkaido salmon scientists and managers are simply not able to enact the ostensibly more cutting-edge wild salmon initiatives prescribed by North American colleagues.[23] Japanese fisheries managers thus struggle to engage with structures of comparison in multiple senses: they must engage both with the comparative framings that since the nineteenth century have firmly positioned Japan as less modern or advanced than the so-called West and with the literal structures of channelized rivers and hatchery fish rearing ponds. Current comparisons from outside Japan, which criticize the state of Hokkaido's rivers and the region's focus on hatchery fish, do not recognize these structures as literally concretized histories of past comparisons—ones in which North American practices of agricultural and fisheries management played a key role.

In the subtle ways illustrated in this chapter, Japanese fisheries professionals often seek to make visible these histories by recounting the comparative origins of Hokkaido's hatchery programs and its broader land/water transformations. In doing so, they enact a comparative practice that is distinct from those of North American fisheries managers, who, by simplistically framing Japanese salmon policy as inadequately attentive to wild fish, ignore how previous comparisons shape the material possibilities of the present. Insisting on such historicity—on how current comparisons unfold in land-water-scapes shaped by past ones—is to insist on imbrication. If North American fisheries managers were to pay attention to such concretized histories, it would indeed shift the grounds of comparison in an important way: rather than evaluating Japanese salmon management from an ostensibly outside position, it would require North American fisheries professionals to reenvision Hokkaido's salmon dilemmas as ones in which their own management histories are also thoroughly entangled.

CHAPTER SEVEN

Other Comparisons

Ainu, Salmon, and Indigenous Rights

I N an English-language overview of Hokkaido's history, author Ann
Irish mentions how Ainu artist Sunazawa Bikky and his brother Kazuo
recounted playing "cowboys and Indians" as children in the 1930s:
"When asked who were the cowboys and who were the Indians, Kazuo
answered, 'nobody wanted to be an Indian, we knew that Indians were
treated the same as us, so we played good cowboys and bad cowboys'" (Irish
2009, 202–3).[1] Far more than mere childhood play, "cowboys and Indians"
enacts a core comparative structure of modern nation-making. Within the
stereotypes of the American Western genre, cowboys are positioned as the
embodiment of a triumphant white masculine American nation, while Indians are positioned as anachronistic, either as vicious warriors whose arrows
are to be defeated by superior guns or as noble savages who are destined to
be crushed by the march of progress.[2] Even as children, the Ainu artist and
his brother were aware of such comparative narratives, which traveled widely
via post–World War II television and film, as well as the related comparisons between ethnic Japanese and Ainu people in Japan. Beginning in the
mid-nineteenth century, Ainu peoples, forcibly enrolled in the comparisons
of Japanese nation-making, were positioned as Japan's Indians, as a constitutive outside for its ostensibly progressive modernity. As the making of a
modern nation-state became an increasingly central concern for the Japanese government, Ainu-ness became something to be eradicated through
assimilation policies. Such "assimilation" was always intended to be partial;
Ainu became Japanese citizens, but unequal ones. From 1899 to 1997, Ainu
were officially classified as *kyūdojin*, or "former natives," a category that at
once denied their indigeneity and blocked them from becoming Japanese.[3]

The Japanese government pursued the colonization of Hokkaido and the assimilation of Ainu people with vigor as part of their own game of cowboys and Indians. In 1879, a *New York Times* article on the ethnology of the Pacific illustrates a comparative interpretation of Japanese settler-colonial history:

> In the earliest records of the Japanese are found accounts of how those "Yankees of the East" landed on the islands they now inhabit, and how they frightened and drove the Ainos from one island to another out of their way, just as, later on, the settlers in this country drove the Indians before them. (*New York Times* 1879)

Such comparisons were clear to Japanese government officials, who wanted to ensure that they became the "Yankees of the East" rather than another set of Indians for the West. They sought to demonstrate their cowboy/Yankee status in part by enacting in Hokkaido what was, in 1879, an already transnationally legible Wild West scene. In subsequent years, such frames further solidified as the Japanese state presented Ainu people as "their savages" at the 1904 St. Louis World's Fair's anthropological pavilion, alongside Sioux, Patagonian, and Pygmy peoples (Carlson 1989; Medak-Saltzman 2010; Vanstone 1993). Contending with the models of modernity proffered by the United States and European nations, the Japanese government chose to forcibly enroll Ainu people in its own comparative regimes, thus thrusting them into history as the Other's Other—the Indians of the East. For Ainu people, there is no easy escape from such regimes of comparison; yet they have never been passive. Violently thrust into a vortex of modern comparison-making, Ainu people both negotiate within and challenge dominant logics by making their own comparisons. These Other comparisons are at once entangled with and in excess of modern binaries.

This chapter focuses specifically on the ways that salmon have been pulled into the Japanese state practices that have usurped Ainu lands, waters, and fish, as well as the ways that Ainu people make comparisons with and through salmon to challenge Japanese state practices.[4] Hokkaido colonization projects sought to secure the island's salmon resources for Japan and exclude Ainu from them as part of efforts to develop an industrial salmon fishery and to force Ainu to become "civilized" farmers. The growing Ainu movements that assert rights to salmon illustrate that while there is no easy escape from the ongoing enactments of Japanese colonial structures and the comparisons entangled with them, there are nonetheless ways to challenge them by making comparisons otherwise.

ENTANGLED WITH SALMON

Archeological remains indicate that people have inhabited the island that the Japanese government now calls Hokkaido—and interacted with its salmon—for at least twenty thousand years (Ono 1999, 32).[5] According to the middens they left behind, most early island inhabitants appear to have eaten at least an occasional salmon, but until the most recent millennia, the inhabitants of this island do not seem to have been salmon-centric. While archeological evidence is always problematic and partial, current research indicates that they hunted large numbers of marine mammals, ate quite a few deer, and farmed barnyard millet and wheat raised from seeds they acquired through trade with Honshu (Yamaura and Ushiro 1999, 45). Similarly, archeological finds indicate that although some of their village sites were located near salmon rivers, many of their communities were located in upland areas away from major salmon spawning grounds (Segawa 2005, 2007). For early inhabitants, salmon seem to have been one species among many—important to be sure, but not indispensable.

But about nine hundred years ago, something seems to have shifted. Villages located on non-salmon-bearing streams seem to have been abruptly abandoned. At the same time, the number of dwellings located near salmon spawning grounds appears to have dramatically increased (Segawa 2007). Suddenly, people could not seem to live without being near salmon. What had changed? Around 1200, the island's peoples established new economic ties with Honshu that transformed their relationships with both salmon and trade goods. Prior to this time, they were clearly involved in significant trade relationships that linked them to the Japanese archipelago, Kamchatka, and mainland Asia. By the tenth century, the island's peoples had already obtained seeds, swords, metal products, and glass (Yamaura and Ushiro 1999, 45). The volume and regularity of such trade, however, seems to have been limited, with imported goods serving as supplements to, rather than replacements for, locally made products (43). Around 1200, at the same time that villages relocated to salmon streams, the number and variety of imported goods—particularly from Honshu—skyrocketed. The influx of goods likely sparked substantial transformations across the island. People appear to have stopped making ceramics as they switched to using imported vessels (45) and to have developed new ritual forms in which Japanese-produced rice and ornate lacquer vessels played central roles (Walker 2001, 112–17).

What were the island's peoples exporting in exchange for all of these new goods? Largely salmon. The island's Ainu peoples found that dried salmon, long a valuable winter food source, were also popular with *wajin* (a term for ethnic Japanese).[6] Inexpensive salmon was a popular protein-rich foodstuff among farmers and other lower-ranking people in northeastern Japan. As Ainu became more entangled in these new economic connections, their relationships with salmon seem to have intensified. They began catching and preserving greater numbers of fish, developing new fishing techniques in the process. They also began to harvest salmon more intensively in river reaches navigable by boat so that they could easily ship the dried fish to distant markets. As salmon became a valued trade good, they also came to take on a larger role in everyday life. Ainu peoples began to eat more salmon themselves, hanging them to dry in the rafters of their houses. They used salmon skin as fabric for making boots, shirts, and children's toys. In short, Ainu peoples became increasingly salmon-centric.[7]

With dried salmon as one of their key products, Ainu peoples extended their already expansive trade networks. Written records from Tosaminato, an important port city along Honshu's Sea of Japan coast, indicate that between 1185 and 1573, Ainu arrived there in their own boats to trade kelp, dried salmon, and sea otter pelts (Kikuchi 1999, 77). But Ainu peoples' trade routes did not link them only to Japan; their trade networks stretched across the Okhotsk Sea and deep into continental Asia. When Japan was allegedly "closed" to the world during the Tokugawa period, Ainu were important transnational brokers who dealt in sea otter pelts from the Kurils, eagle feathers from Kamchatka, and fabrics from China (Segawa 2007). The people and landscapes that emerged from such exchanges were highly cosmopolitan; Ainu engaged with Aleuts, Indigenous Kamchatkans, Russians, and Mongolians. Furthermore, prior to *wajin* colonization, Ainu were already farming crops that originated in the Western hemisphere, including potatoes and two types of American squash (Kohara 1999, 204–5).

In the sixteenth and early seventeenth centuries, however, challenges began to mount for Ainu. In 1604, the Tokugawa shogunate granted one of its feudal domains, the Matsumae han, a charter that gave them exclusive rights to trade with the Ainu (Siddle 1999, 69). The Matsumae domain invited traders from Honshu to set up offices at the southern tip of Hokkaido and work as their agents, bringing profits to the feudal domain's coffers. Matsumae traders took advantage of their monopoly—backed up by substantial military might—to exploit Ainu peoples. First they blocked Ainu people from traveling to Honshu to trade on their own terms. According to one scholar, "After 1644, Ainu boats were no longer to be seen in Tohoku [northern Honshu] ports, an indication of the success of Matsumae attempts to monopolize trade" (Siddle 1999, 69). Matsumae traders were highly exploitative; seeking to maximize their profits, they significantly reduced the amount of rice that they paid Ainu for dried salmon. Ainu peoples protested the unfavorable rates of exchange, eventually waging a war against the Matsumae domain in 1669. For Ainu, the goal of this conflict (called Shakushain's War) was not to entirely sever relations with *wajin* but to end the Matsumae domain's monopoly and return to more just trade relations (Howell 1999, 97). The Ainu were militarily defeated, and afterward, they became subject to progressively more exploitive Japanese demands.[8]

In the early eighteenth century, the Matsumae domain began subcontracting trading posts, located along the coast of Hokkaido, directly to Honshu traders.[9] Although the Matsumae continued some trading with Ainu peoples, these posts also developed a system called *basho ukeoi*, where *wajin* subcontractors brought in their own boats and nets, repositioning Ainu as laborers, not trading partners. The *wajin* traders cornered Ainu into this direct-labor system through violence and threats of violence. Sometimes, ethnic Japanese traders relocated entire Ainu villages to camps next to trading posts. At other times, they rounded up Ainu men and shipped them to distant parts of Hokkaido to labor in the fisheries there (Walker 1999, 103). In 1858, a Japanese official noted that "of forty-one Japanese fisheries supervisors in Kushiro, thirty-six had taken Ainu women as 'concubines' after sending their husbands to work at the neighboring Akkeshi fishery" (103). While many men were forced to labor in Japanese salmon fisheries, women, children, and the elderly struggled to catch and preserve enough salmon for subsistence use and trade. Furthermore, intensified interactions with the fishing posts brought Ainu people into contact with new diseases, including smallpox and syphilis, that further affected their communities. Recurring epidemics were documented from the seventeenth century onward (Walker 2001, 181). In 1807, shogunal officials recorded 26,256 Hokkaido Ainu; forty-seven years later, they tallied 17,810. Regional-level population estimates point toward even sharper declines and devastating effects (182).

The introduction of ethnic Japanese salmon harvesting also impacted the fish. Ainu peoples typically harvested the majority of their salmon at or near the fishes' spawning sites, after they had laid their eggs and released their milt. Because most of these salmon had already reproduced and were on the verge of death, one could harvest a large number of such fish without endangering future generations of salmon. In addition, Ainu people chose these fish because post-spawning salmon made for longer-lasting dried salmon. Because salmon consume most of their fat reserves as they produce gonads, migrate upstream, and dig their spawning nests, post-spawning salmon are exceptionally lean. Salmon caught in the ocean had such a high fat content that they could not be effectively dried; they would spoil too quickly. Post-spawning salmon, however, had nonoily flesh that could be easily dried and that could last more than a year without becoming rancid. The duration that the salmon remained edible mattered, as Ainu people could then trade these long-lasting fish in the spring following their harvest, a time when ocean waters were much calmer and allowed for safer boat travel.[10] Ethnic Japanese, however, harvested salmon in a different way. They typically caught salmon in bays or at river mouths, long before the fish reached their spawning grounds. The salmon not only did not have a chance to reproduce before capture; they also had a very high fat content. Their oiliness required a different kind of processing, one that involved large quantities of salt. Ethnic Japanese transported salt from Honshu to their remote Matsumae trading posts to sustain their salmon industry.[11]

Ainu people were forced into a corner; they had fewer salmon at their upriver fishing sites, as more fish were harvested at river mouths by the Japanese, and they had fewer people to harvest them, as more of their men were forced to labor at Japanese fishing stations, yet they still needed trade goods beyond the minimal rice that Ainu men received in exchange for their work. Furthermore, Honshu residents tended to prefer *wajin*-style salted salmon to Ainu-style dried ones. Regardless of processing method, prior to the twentieth century, Ainu Mosir's salmon were a staple protein source for poor Tohoku farmers and lower-class urban residents rather than a "fish of kings."[12] By the time they reached markets, salmon from Ainu Mosir were as hard as rocks, so tough that they could not be cut with a knife. To eat the salmon, one had to first soak it in water or broth. While such problems ran across all processing methods, ethnic Japanese style salted salmon was a bit softer and thus considered a bit higher quality. Ainu salmon producers, however, did not have access to the salt resources needed to produce that form of preserved salmon. Unable to compete with the salted fish, they were forced to sell their unsalted dried and smoked fish for prices lower than those fetched by salted ones. These ethnic Japanese forms of salmon processing also impaired the salmon reproduction, as their mode of producing fattier salted fish shifted the harvest to pre-spawning fish. Salmon had thrived within Ainu worlds, where people caught them largely after they had laid their eggs, but it was difficult for them to do so within ethnic Japanese ones.

Within the *basho ukeoi* system, Japanese traders and merchants sought to enroll Ainu in unequal economic relations, but they did not engage in projects that explicitly aimed to craft Ainu identities.[13] Their goal was to produce profit, not citizens or state territory. After the formation of the Meiji state, however, the goals of ethnic Japanese engagements with the Ainu and Ainu Mosir shifted from commercial exploitation to governance. With Western imperial nation-states as their model, the central Japanese government wanted the island to be more than a place within a loose Japanese sphere of economic influence; they wanted it to be specifically Japanese territory, to lie within the body of the nation. With this new project, the Japanese state was no longer content to exploit Ainu people; they now wanted to make them into national subjects. Beginning in 1869, the central Japanese government began its campaign to make the island "Japanese" by aggressively promoting both ethnic Japanese settlement of Ainu Mosir and Ainu assimilation.[14]

Desires for resource exploitation did not require such changes. The *basho ukeoi* system did a brutally outstanding job of extracting salmon (as well as herring) from the island's waters.[15] Because Ainu laborers were able to partially feed themselves through gathering, hunting, and fishing, ethnic Japanese could compensate Ainu at a level below what was necessary to sustain them as laborers. In contrast, the Meiji state was more concerned about forceful claims to territorial sovereignty, fearing that if they did not assert rule over Ainu Mosir, Russia soon would. To bring Ainu Mosir into the fold of the Japanese nation, they sought to make it undeniably Japanese.

At the same time that the Japanese government promoted ethnic Japanese migration to Hokkaido, they also sought to slot Ainu people into their projects of nation-making. The Japanese government wanted the Ainu to become Japanese state subjects, but in a way that positioned them as marginal and lesser than ethnic Japanese. While the Japanese government wanted to make the Ainu "Japanese" so that they did not become Russian, they also positioned Ainu as second-class citizens to justify the colonization of Ainu land. Historian Tessa Morris-Suzuki describes how such paradoxical goals shaped citizenship in Japan's colonies: "The ruling state's urge to exalt and spread the values of its own 'civilization' contended with its desire to maintain the differences that justified unequal access to power" (1998a, 161). While the government sought to assimilate the Ainu, they actively pursued measures, including special land policies and financial controls, that ensured that the Ainu were "assimilated" as relatively power-less, impoverished citizens.

Efforts to remake Ainu identities became a clear state project, one that the Japanese government enacted through countless comparisons. Imme-diately after annexing Hokkaido, the Japanese government banned the *basho ukeoi* system and "freed" the Ainu from forced labor. They then turned to the American West as they considered how to fashion the Ainu into citi-zens. In dealing with the so-called Ainu problem, Hokkaido colonial offi-cials drew on a particular strain of US Indian policy—that which stressed assimilation over reservations. As noted in chapter 2, they solicited the opin-ions of Horace Capron, one of the American advisors to the Kaitakushi, who had previously served as a US government Indian agent in Texas (Medak-Saltzman 2008, 97). Capron was an enthusiastic supporter of US efforts to convert Indians into farmers; he was also a proponent of the 1877 Dawes Act, which broke communal Indian lands into individual allotments for native families (freeing up "excess" lands for white settlers) (102, 104). In building their own policies, Hokkaido officials drew on Capron's opinions as well as on US institutional forms. Japanese leaders, such as Nitobe Inazō (an instructor at the Sapporo Agricultural College and a government offi-cial), was familiar with the native policies of New Zealand and other coun-tries but seemed particularly inspired by the Dawes Act and comparisons with US Indian policies, personally translating into Japanese an American's 1894 speech about the act (Harrison 2009, 99).[16]

But there were also other comparisons at play. Japanese government officials were also comparing the Ainu with themselves. In the bizarre worlds where the status of "colonizer" marks a nation as "civilized," Japa-nese officials sought to prove that they were building a modern nation by constructing the Ainu as their inverse—as people to be colonized. Through brute force, unjust policies, and narratives of Ainu "primitive-ness," the Japanese state attempted to convert the Ainu—a prosperous and worldly trade society—into a "dying" culture in "need" of colonial uplift. Erasing histories of violence, the Japanese government turned Ainu assimilation policies into an imperative to uplift poor primitives, helping them to achieve a more civilized form. Although the primitive/civilized dichotomy is common across state-making endeavors, the Japanese state had its own civilizational ideals. In the Meiji period, notions of Japanese-ness were tied to a very specific multispecies formation: rice paddy agri-culture. As anthropologist Emiko Ohnuki-Tierney (1993) has described, ethnic Japanese have consistently used rice to negotiate boundaries between self and other. Claiming "rice as *our* food" and "rice paddies as *our* land," ethnic Japanese have defined themselves as fundamentally "agrarian" (regardless of the actual occupations of most Japanese) (4). Within this logic, making a landscape Japanese has meant the "transformation of wilderness into a land filled with succulent heads of rice. In short, rice paddies created 'Japanese land'"(132). From an ethnic Japanese state perspective, then, the ideal way to make the Ainu "Japanese" would have been to turn them into rice farmers. However, until the development of cold-resistant rice strains in the 1930s, rice cultivation was difficult in Hokkaido (Irish 2009, 220). In the face of long, frigid winters, the Hokkaido government decided to try to convert Ainu people into farmers, but with wheat, corn, sugar beets, and beans as substitutes for rice.

Ethnic Japanese traders had already secured access to Ainu fisheries, but in order to force Ainu into farming, they sought to outright prohibit their ability to fish. As long as they possibly could, Ainu people sought to maintain their own ways of life, preferring salmon fishing, hunting, foraging, and farming on their own terms to the agricultural lots assigned by the Meiji government. In the late nineteenth century, the Japanese government recognized that as long as the Ainu had continued access to salmon and other resources, they were not likely to accede to state plans. As a result, the Hokkaido Colonization Commission sought to eliminate Ainu access to and relationships with salmon. In 1877, the commission established a hatchery beside an Ainu village on the Chitose River at the same time that it banned fishing in that river basin (Kosaka 2018, 71). While Sapporo-based officials raised concerns about depriving Ainu people of salmon, nineteenth-century government officials in Tokyo insisted that plans to combine hatcheries and fishing bans go forward, replying to the Sapporo office:

> We expect artificial breeding will bring about economic benefit
> in the future. When you take total gains and losses into
> account, the damage to the minority can be ignored. You should
> not adhere to residents' welfare. They may be driven to be
> farmers. (Yamada 2011, 168, as translated in Kosaka 2018, 71)

In 1879, the Colonization Commission banned salmon fishing in more of Hokkaido's rivers, claiming that such an act was necessary to protect the island's salmon populations from overharvest (Aoyama 2012, 119). The ban, however, was a barely veiled attempt to eliminate Ainu lifeways, and it did nothing to conserve fish. Because Japanese commercial fishermen harvested salmon in the ocean and in the mouths of rivers, rather than in the rivers themselves, the new freshwater salmon fishing ban had no effect on their activities. The Japanese fishermen continued to harvest huge numbers of salmon with abandon, while all Ainu fishing was rendered illegal. Ainu people had no access to the capital necessary for large coastal fisheries operations, and they were completely dependent on upriver fisheries, where they could harvest easy-to-preserve low-oil fish. The Japanese claim that river-harvest bans were necessary to preserve salmon spawning was a ruse; because Ainu people typically harvested salmon after they spawned, their fishing activities had minimal impacts on salmon populations. In reality, the intent of such laws was to force Ainu people to stay on government-assigned plots and to participate in assimilation programs.

Without access to salmon, the Japanese government realized, Ainu people could not be Ainu. In his memoir, *Our Land Was Once Forest*, prominent Ainu activist Kayano Shigeru (1926–2006) wrote that the "law banning salmon fishing was as good as telling the Ainu, who had always lived on salmon, to die. For our people, this was an evil law akin to striking to death a parent bird carrying food to its unfledged babies" (Kayano 1994, 58–59). After the ban, Ainu people tried to continue salmon fishing, but they became "poachers" in their own rivers. The Japanese government began to crack down on Ainu salmon fishermen, arresting them and putting them in jail. Ainu people who tried to remain Ainu—who tried to feed their families with salmon—became "criminals" (57–61).[17]

Beginning in the late nineteenth century, the Japanese government also used hatcheries to further disassociate salmon from the rivers and Ainu communities. The Meiji era government's campaign to move salmon spawning out of rivers and into hatcheries radically remade Hokkaido's salmon and watersheds, modified salmon genetic population structures, and altered regional ecologies by removing the nutrient inputs that salmon carcasses provide. But moving salmon into hatcheries also aided policies aimed at assimilating Ainu people. Hatcheries made enforcement of river fishing bans easy; they virtually eliminated salmon from Hokkaido's rivers. Hatcheries used weirs to block upstream salmon migrations, capturing brood stock for their programs near the mouth of rivers. With the advent of hatcheries, Hokkaido's Japanese commercial fishermen no longer needed rivers and their salmon spawning grounds to fill their nets. Ainu people could do little but watch the numbers of salmon spawning in Hokkaido's rivers plummet as more and more waterways were used for hatchery production, blocked by dams, or degraded by channelization and pollution.

These changes not only damaged Hokkaido's environment; they also fractured the multispecies relationships at the core of Ainu worlds. For example, Blakiston's fish owls (*Bubo blakistoni*) are gods who guard Ainu villages, depend on healthy river habitats, and feed on salmon carcasses. Although they once ranged across Ainu Mosir, the owls, critically impacted by river modifications and the hatcheries that relocated salmon bodies and their nutrients to human food and fertilizer industries, have been listed as an internationally recognized endangered species (Japan Bird Research Association 2010). The direct losses from salmon industrialization were amplified by terrestrial colonization and development initiatives, including systematic, government-sponsored hunting of more than half a million Ezo deer between 1873 and 1878 alone. The resulting venison was canned and exported, drawing on the same techniques used in salmon canneries and often targeting similar export markets in France and the United States (Hirano 2015, 205–6; see also B. Walker 2005, 148–50). Wolves were also exterminated as part of projects to establish safe pasture for cattle, horses, and sheep as timber harvests deforested large tracts of land (B. Walker 2005).[18] These acts to make Hokkaido more "productive" at once upended ecological assemblages and killed Ainu people; according to one estimate, between the 1870s and 1920s, more than 70 percent of the island's Ainu population died, often from starvation (Hirano 2015, 214).[19]

In the century after salmon were forced into hatcheries and Ainu people onto farms (and into starvation), Ainu, as "former natives" stuck in a limbo produced by the Japanese imperial nation-state, were forced to assimilate but denied the opportunities to actually do so, as racial prejudices often blocked their efforts to pursue educational opportunities or obtain mainstream jobs. In the face of such challenges, Ainu-ness was sometimes transformed, sometimes forgotten, and sometimes actively expunged. Many Ainu people hid their identity, adopting Japanese customs and speaking only Japanese. They often did not tell their children about their Ainu heritage to try to spare them the stigma of being Ainu. One acquaintance of mine who suspects that she may be of Ainu decent said her now-deceased parents refused to tell her anything about her grandparents—even their names. Within such contexts, Ainu peoples' relationships with salmon did not disappear, but they significantly changed. During the twentieth century, in lieu of salmon fishing and bear hunting (which had become difficult to enact), activities such as dance, song, clothing, and art became more common enactments of Ainu-ness. Some Ainu people were able to garner commercial fishing rights with Japanese government systems by becoming members of fishing cooperatives and applying for salmon set net licenses (chapter 5). Yet in these settings, salmon were a market product, and overt Ainu-ness was not welcome. While constrained through dispersed social discrimination as well as targeted legal maneuvers, Ainu relations to salmon remained present enough to become a central part of the efforts—beginning in the 1970s and intensifying in recent decades—that have been variously referred to as Ainu revitalization, revival, resurgence, and efflorescence (Roche, Maruyama, and Kroik 2018; Uzawa 2018).

OTHER COMPARISONS

An Ainu man once asked me if I had heard about how the Japanese divide the world into two kinds of people—rice people (Japanese) and bread people (Westerners). It is wrong, he told me. The world as he saw it had three—not two—kinds of people: rice people, bread people, and salmon people. The Ainu, he explained to me, occupied a third space. At the same time that he implicitly accepted certain comparative premises—including the seemingly natural juxtaposition of Japan and the West—he was also seeking to undo the binarism that underpins them and makes a space for another way of being.

Throughout experiences of colonial violence, Ainu people have always been making Other comparisons—comparisons that are at once engaged with and distinct from those made by the modernist Japanese state. Their Other comparisons do not come after but are rather contemporaneous with those of Japanese nation-making. At the 1904 World's Fair, mentioned earlier, Ainu were not passive "objects" displayed by the Japanese; instead, they interacted with other Indigenous peoples (Medak-Saltzman 2010). One photograph from the 1904 fair documents a meeting between an Ainu woman named Santukno Hiramura and a Patagonian Tzoneca woman named Lorenza (592). As she bent curiously toward Lorenza, who was holding her dog named Kik, Santukno Hiramura was likely enacting other comparisons within and against the modern/primitive ones that underpinned the 1904 World's Fair (596).

The importance of transnational comparisons and alliances for Ainu movements is more clearly documented from the 1970s onward. Interactions and comparisons with other Indigenous peoples have assisted Ainu to develop their own modes of challenging myths of Japanese homogeneity and an intransigent Japanese state, which repeatedly denies their rights claims (Siddle 1996, 2). For example, in a 1977 newspaper report of a meeting between Ainu leaders and two Inuit representatives, one of the Inuit makes clear the comparability of their claims, stating that "the Ainu have their rights and they are the same rights as those of the Eskimo" (quoted in Larson et al. 2008, 58). In the early 1980s, Ainu leaders began regularly participating in international conferences, such as the World Council of Indigenous

Peoples (Larson et al. 2008, 58), and in 1992, Giichi Nomura, the executive director of the Ainu Utari Association, was invited to give the opening address at the United Nation's launch event for its International Year of the World's Indigenous People.[20]

As Ainu people develop their own forms of Indigenous identity, cultural resurgence, and rights movements, such comparisons produce neither certainty nor solidity but open questions. Like any Indigenous or ethnic group, Ainu people have diverse opinions about what it means to enact Ainu-ness. Furthermore, the thousands of Ainu who live in Tokyo often describe different experiences of Ainu identity than those who reside in Hokkaido (Uzawa 2020; Watson 2014). For some, being Ainu is primarily about *bunka* (culture)—song, dance, handicrafts, and festivals. Others place a stronger emphasis on the struggle for recognition of Ainu *kenri* (rights). For many younger Ainu whose parents hid their ancestry in the midst of discrimination, shifting relationships to their own Ainu-ness is not uncommon; some describe being Ainu at certain times and in certain contexts but not others.[21] When I was trying to establish contact with Ainu commercial fishermen, an Ainu man told me not to bother looking for them at the local fishing cooperative. In the context of the fishing cooperative, no one is Ainu, he told me. If I want to find Ainu fishermen, I needed to go to local Ainu events and ask who fishes commercially.[22]

In the midst of this diversity and exploration, salmon are integral to a wide range of Ainu resurgence efforts. Intertwined with cultural forms such as the first salmon ceremony but also linked to issues of natural resource access, salmon swim at the interface of culture- and rights-focused modes of enacting Ainu-ness. Through public *ashiri chep nomi*, or first salmon ceremonies, Ainu people seek both to foster Ainu community and to increase visibility within spheres dominated by assumptions of Japanese homogeneity.[23] Especially since the passage of the Ainu Cultural Promotion Act in 1997, various city governments have been supportive of these festivals as displays of Ainu culture.[24] But these ceremonies also implicitly challenge the idea of "culture" as separable from rights as the ceremonies require access to salmon. In their materiality, salmon create a slippage between *bunka* (culture) and *kenri* (rights) that is important to Ainu movements, as well as to the salmon themselves.

After years of Ainu advocacy, the Hokkaido government began to allow very limited and circumscribed Ainu salmon harvests in 2000 under the rubric of "cultural promotion." Within this frame, the Hokkaido prefectural government allowed limited Ainu-style salmon harvests not through an idiom of Indigenous rights but through languages of historical preservation and cultural revitalization, carefully worded to avoid legally acknowledging any Ainu claims to fish.[25] The special harvest permits, granted for specific rivers and time frames, typically allow for the harvest of only a small number of fish—often three to eight fish a person or fifty to one hundred for a group ceremony—with stipulations that prevent any of the salmon from being sold. As one Ainu man explained to me, the Ainu community of which he is a part initially asked the government to allow them to catch a few fish simply because they wanted their ceremonies to incorporate salmon caught with traditional tools rather purchased from a grocery store.

Yet such access to salmon, even in this limited form, has brought Ainu people into deeper conversation with transnational Indigenous rights movements and sparked more expansive rights claims. Such dynamics are visible within an attempt to block the construction of an industrial waste disposal site along a salmon-bearing river in northern Hokkaido. In this movement, which began in 2009, Ainu people, urban environmentalists, and local residents came together to obtain the first pollution control agreement that protects wild salmon and recognizes Ainu rights. The agreement itself is significant, but it is not the only outcome of opposition to the waste dump. Through this effort, the region's Ainu community, Ainu rights discourses, salmon conservation policies, and the evolutionary trajectories of local salmon have taken new directions.

MONBETSU AINU

Satoshi Hatakeyama-ekashi is the head of the Ainu organization in Monbetsu, a town perched on the edge of the Okhotsk Sea, and its most vocal member.[26] Indeed, Hatakeyama-ekashi takes up so much space that it often seems that he is the entirety of the Monbetsu Ainu branch. In his late sixties, Hatakeyama-ekashi, with his thick neck, square jawline, and booming voice, is the region's most visible and outspoken Ainu. Growing up in Monbetsu, a rural area known for its fishing and dairy industries, Hatakeyama-ekashi always felt marginalized. A descendant of a local Ainu leader who governed several small villages in the late nineteenth century, Hatakeyama-ekashi was born in an Ainu *kotan* (settlement). Because everyone knew about his family background, he faced such serious bullying as a child that he dropped out of school before completing the seventh grade. While Hatakeyama-ekashi sometimes struggles to read kanji characters, he is a smart and savvy businessman who owns his own commercial fishing boat. After facing discrimination in his youth, he spent most of his adult years trying to distance himself from his Ainu heritage, refusing to attend Ainu festivals or related events. Occasionally, his Ainu heritage continued to dog him. For example, when Hatakeyama-ekashi became a fisherman, the local fishing cooperative initially refused to admit him as a full member, relenting only after Hatakeyama-ekashi had an official from the Ainu Association's Sapporo headquarters pressure the co-op to drop their discriminatory stance. For decades, Hatakeyama-ekashi did his best to hide his Ainu-ness, to be as Japanese as possible. But about thirteen years ago, he decided to publicly express his Ainu identity after his older brother died. Hatakeyama-ekashi's brother had embraced their Ainu heritage, attending festivals in other towns with more active Ainu communities and making a deathbed request for an Ainu funeral. But even those gestures did not convince Hatakeyama-ekashi to return to the Ainu fold. After his brother passed on, Hatakeyama-ekashi decided to "quit being Ainu" once and for all (*Ainu wo yameru*). But Hatakeyama-ekashi's deceased brother objected to this plan. He visited Hatakeyama-ekashi in a dream, urging his younger brother to reclaim his Ainu-ness.

Hatakeyama-ekashi decided that a request from the other world was not to be ignored. In 2002, he rekindled the local Ainu branch and began organizing *ashiri chep nomi*, or first salmon ceremonies, in Monbetsu. But although Hatakeyama-ekashi knew quite a bit about hiding Ainu-ness, he found that he knew little about how to more visibly enact it. He did not know any prayers or songs or how to use an Ainu fish spear, called a *malek*. When Hatakeyama-ekashi wanted to hold a first salmon ceremony, he invited Ainu elders from other parts of Hokkaido to lead the event because neither he nor anyone else in Monbetsu knew how. Such a situation is not uncommon. In the wake of intensive assimilation pressures, Ainu often turn to each other and to ethnic Japanese scholars to revitalize various practices. But while many Ainu try to base their cultural practices on oral histories with elders and carefully researched historical data, Hatakeyama-ekashi was less interested in questions of cultural authenticity and more focused on issues of rights. Hatakeyama-ekashi scheduled the 2010 Monbetsu first salmon ceremony for August instead of in September (as is common for other Ainu groups) so that it would coincide with an environmental event hosted by ethnic Japanese activists, who seemed like potential allies for his efforts to block the construction of a waste disposal site in the upper reaches of the Monbetsu watershed by asserting his Indigenous rights. Hatakeyama-ekashi was undaunted by the fact that in August, there are almost no chum salmon in its waters, only pink salmon, which are considered trout within Ainu and Japanese classificatory systems. Hatakeyama-ekashi simply turned the first salmon ceremony into a first trout ceremony, with a different silver fish on the ritual altar. When only a few Ainu people showed up for the unseasonal event, Hatakeyama-ekashi did not hesitate to draft non-Ainu—including this anthropologist—to fill ceremonial roles, as the ritual itself became a space for new kinds of multiethnic collaborations.[27]

Hatakeyama-ekashi and his group are not only geographically far from Hokkaido's larger Ainu communities; they also often work outside of the established channels of the Ainu Association of Hokkaido. Although officially an independent entity, the Ainu Association receives government funds for cultural revitalization activities, and its leaders tend to be more restrained in the demands that they make on the Japanese state. In contrast, Hatakeyama-ekashi forges his networks largely through collaborations with ethnic Japanese NGOs in Sapporo and Tokyo that focus on social justice, transnational Indigenous rights, and environmental protection. These alliances are not without tensions, negotiations, and compromises. Hatakeyama-ekashi is vocally pro-whaling—not only as an Indigenous but also as an industrial practice—and for a short time, he worked on one of the Japanese whaling ships dispatched to Antarctica, which are opposed by most international environmental groups. Furthermore, in his early days of collaborating with the environmental and social justice groups, Hatakeyama-ekashi also suggested that the Ainu establish their own high-value salmon hatcheries by introducing Chinook and coho from the United States, an idea that runs counter to notions of biodiversity conservation.

MONBETSU SALMON

Yet despite such differences, Hatakeyama-ekashi was able to make Monbetsu's Ainu-salmon worlds of interest to such groups. This took substantial work, as the town's river system and its salmon were not, on their own, seen as ecologically valuable. With a fifteen-foot-high concrete slab embankment on one side and a thirty-foot hill of bare dredge spoils and a mix of gravel and broken scallop shells on the other, the mouth of the Monbetsu River is far from being an exemplar of romantic nature. One of the upper tributaries where adult salmon spawn is a straight, four-foot-wide agricultural drainage ditch, its banks lined with pasture grasses, while another section flows outward from a denuded construction site and over an earthen dam covered by a blue tarp.

While international salmon and river conservation groups have taken an interest in some of Japan's aquatic worlds, the Monbetsu is not one of them. When fisheries professionals seek to protect "wild salmon," they are typically seeking to conserve genetic specificity, which the Monbetsu salmon are thought to lack. Although the exact history of the river's fish is unknown, records indicate that hatchery-reared juvenile salmon were trucked to the river and released into it prior to 1994. The river's current fish are probably descendants of these earlier fish releases, along with more recent strays from nearby hatcheries. While Monbetsu salmon are likely "wild" under Hokkaido law, which defines fish as such once they have spawned outside a hatchery for at least two generations, fish with such recent hatchery backgrounds are not considered fully wild by many salmon biologists (especially those outside of Japan), because they do not have a specific genetic link to their river or adaptations that make them a distinct population from those in nearby hatcheries. In the comparisons of major international salmon and river conservation groups, Monbetsu salmon are seen as less valuable than those whose genetics, behaviors, and populations have been less affected by hatchery practices. When a North American environmental group decided to invest in salmonid conservation in northern Hokkaido, they were drawn to a less concretized and more scenic river to the west of Monbetsu with a population of Sakhalin taimen (*Parahucho perryi*), a species closely related to salmon that has never been subject to artificial propagation. Listed as critically endangered by the International Union for Conservation of Nature, the taimen and their river more closely aligned with established environmental priorities.

WASTE DISPOSAL SITE STRUGGLE

However, through Hatakeyama-ekashi's efforts, the Monbetsu Ainu community and the Monbetsu's more-than-human assemblages began to garner more attention. In June 2008, just days before Hokkaido played host to that year's G8 Summit, the Japanese government announced that it would officially recognize the Ainu as Indigenous people. Because Japan had already signed the UN Declaration on the Rights of Indigenous Peoples the previous year, this meant that in principle, the Japanese government would be bound by international law to recognize Ainu rights. But in the months following this official recognition, nothing changed. The Japanese government set up a committee to "study" Ainu issues, taking no immediate actions and making no changes to domestic laws. Hatakeyama-ekashi soon became frustrated by what increasingly seemed to be a meaningless gesture. He wanted real "rights recovery" (*kenrikaifuku*). He began sending formal letters to the Hokkaido governor, petitioning the Hokkaido prefectural government to live up to the central government's announcement and recognize Ainu rights. At first, Hatakeyama-ekashi cast his net widely, making broad appeals for scholarships for Ainu youth, Ainu participation in natural resource management, Indigenous fishing and whaling rights, and economic empowerment programs. Hatakeyama-ekashi's primary goal was to assert that Indigenous rights cannot merely be enacted through empty and abstract words but must be meaningful in the everyday lives of Ainu people. The Hokkaido government did not respond.

Hatakeyama-ekashi was also facing another dilemma: the impending construction of a waste disposal site in the upper reaches of the river where the Ainu group harvested its ceremonial salmon. In 2005, the local government had decided to stop accepting industrial waste at its public municipal landfill in order to extend its life. The decision proved costly for the town's processing plants and agricultural firms, which had to pay to transport and dispose their waste outside of the city. Groups such as the food manufacturer's association, famers' union, and dairy union lobbied city officials for a new facility. They soon drew up plans for a forty-one-hectare repository on one of the hillslopes of the Monbetsu River watershed (Noguchi 2017, 205). In 2007, Hokkaido Prefecture approved the proposal.

From the beginning, Hatakeyama-ekashi strongly opposed the project as both a fisherman and an Ainu. The Monbetsu watershed was already a mess. Its mixed conifer and broadleaf forests had been heavily logged in the mid-twentieth century and replanted with a non-native pine species. Its waters had been polluted by an upstream gold mine that continued to leach chemicals into the river. In Hatakeyama-ekashi's opinion, the watershed did not deserve any more insults. One tactic for stopping this kind of project could have been to mobilize the fishermen's cooperative. In Japanese environmental politics, fisheries cooperatives, which have recognized stakes in maintaining water quality, have played important roles in demanding improved pollution control and resisting the construction of seaside nuclear power plants. But this time, the co-op was not on the side of the Monbetsu River and its salmon. The river itself produced few of the salmon that the cooperative harvested, most of which came from a hatchery on another river. Furthermore, the fishing co-op had been enrolled as one of the alleged beneficiaries of the new waste dump, where the town's seafood processors would be able to deposit scallop shells, a by-product of the fishing co-op's most valuable species. Because the co-op needed another place to put shells, it did not oppose construction of the disposal site, despite the project's potential to leach dangerous chemical compounds into local waters.

Hatakeyama-ekashi, frustrated by the fishing cooperative's shortsightedness, began to wonder if he could use his Ainu-ness to block construction of the waste dump while simultaneously advancing Ainu rights. "I'm not doing this *as* Ainu *for* Ainu," he once told me about his anti-dump efforts. His aim was to show that Ainu rights could be used to protect the environment and benefit the larger Monbetsu community, that Ainu rights were not about taking resources from others or asking for handouts but about using rights to give back and enrich the town. The city council, however, was not impressed when Hatakeyama-ekashi began arguing that the UN Declaration on the Rights of Indigenous Peoples required that they consult with him before ruling on the waste dump's construction permit. Ignoring his demands, the city council approved the final construction permit in February 2010 without consultation with local Ainu.

Frustrated and angry, Hatakeyama-ekashi turned to the alliances that had already been supporting his efforts for several years. Since 2008, Hatakeyama-ekashi had collaborated with a community educator and head of a Japanese "freedom school," a social-justice-education NGO loosely inspired by the Freedom Schools of the American civil rights movement. In 2009, the school organized a study tour, in which I participated, that helped spread awareness of his fight to block the waste dump and connect him with a representative from the Japan Council on the UN Decade of Education for Sustainable Development, an umbrella networking organization for Japanese NGOs with links to UN programs. Those allies and their onward connections allowed Hatakeyama-ekashi and his Monbetsu group to submit an official statement to the UN Human Rights Council and amass fifty-six documents of support from international Indigenous organizations (Noguchi 2017, 208–9). Hewing to Japanese colonial logics—birthed within a particular set of comparative practices—the national government, Hokkaido Prefecture, and other Japanese administrative units continue to refuse Ainu rights while insisting on the legitimacy and primacy of the nation-state's territorial sovereignty. To counter such comparisons, Hatakeyama-ekashi and his NGO allies mobilize their own: those associated with Indigenous rights. By comparing Japan's ongoing refusal of Ainu rights to international standards for Indigenous recognition, rights, and environmental comanagement, they try to portray as out of date the Japanese government's reticence to acknowledge the settler-colonial violence it enacts in the name of nation-state modernity. By comparing Japan's Ainu policies to Indigenous rights legislation in places such as New Zealand, Canada, and Scandinavia, the Ainu coalition seeks to shift the Japanese government's stance on the waste disposal site processes through comparative pressures—especially the shame of falling behind in international arenas.

Such efforts indeed forced some changes in Hokkaido governance practices, including the admission of Hatakeyama-ekashi and the Monbetsu chapter of the Hokkaido Ainu Association into the Hokkaido Industrial Pollution Examination Panel's arbitration proceedings in relation to the waste disposal site. As part of that process, the panel formally recognized the Monbetsu Ainu as a stakeholder, yet to the disappointment of Hatakeyama-ekashi's coalition, the panel also ruled that the construction and operation of the dump could proceed. Still, based on the stakeholder recognition, in March 2012, Hatakeyama-ekashi's alliance pressured the company building the waste disposal facility into signing a pollution control agreement (*kōgai bōshi kyōtei*) directly with the Monbetsu Ainu group, acknowledging the rights of the Monbetsu Ainu to inspect the operation at any time and to receive regular monitoring reports (Noguchi 2017, 208, 209). The agreement was legally significant; until the Monbetsu case, only local government authorities had been considered legitimate signatories of pollution control agreements. But it did not stop the facility from being built.[28]

COMPARING WITH SALMON

Salmon were themselves significant to these alter-comparative practices. Hatakeyama-ekashi's claims to rights and stakeholder status were made possible by the Monbetsu salmon and the very ceremonial fish harvests laws that had been written to obviate Ainu rights claims. At the time of the waste dump controversy, Hatakeyama-ekashi and the Monbetsu Ainu group had applied for and received permits for "cultural promotion" salmon harvests in the Monbetsu River for about a decade. As previously mentioned, the Hokkaido government carefully crafted the law so that it does not acknowledge Indigenous rights or even ethnic difference. The law is written such that even as a foreigner, I was able to be part of an application for one of its permits to harvest salmon for "cultural purposes." But Hatakeyama-ekashi torqued this law that was designed to be legally impotent to make nascent rights claims, a move aided by the legibility of salmon fishing activities within transnational Indigenous rights spaces.

Salmon rights struggles have a deep history in US and Canadian histories of Indigenous activism. During the 1960s and 1970s in the US Pacific Northwest, and in the Columbia River in particular, salmon "fish-ins" were a central practice for asserting tribal rights. At such events, American Indians refused to buy state fishing licenses to catch salmon at off-reservation traditional harvest sites to assert their ongoing treaty rights to fish—rights that US state and federal officials willfully ignored. The American Indian activists then used their arrests and fines to bring court cases through which they successfully argued for their legal rights to fish.[29] In subsequent decades,

Columbia River tribes have secured more substantial participation in salmon management and restoration by organizing through the Columbia River Inter-Tribal Fish Commission, which they established in 1977. Such histories of salmon-related Indigenous activism are not unique to the Columbia River but instead stretch from California through Alaska, a region where diverse Indigenous communities share deep relationships to salmon and most groups practiced some form of a first salmon ceremony. Through their *kamui chep nomi*, Hatakeyama-ekashi and the Monbetsu Ainu group emphasized their kinship with and comparability to other North Pacific salmon people, whose rights to fish have been acknowledged to a greater degree than they have been in Japan. In this way, salmon embodied and fostered an important comparative frame.

Hatakeyama-ekashi's comparative rights practices were also dependent on the materiality of the Monbetsu salmon. Without their bodily presence in the river, it would have been more difficult for him to assert his rights to participate in the waste disposal site proceedings, draw the attention of international Indigenous organizations, and even garner a ceremonial fishing permit in the first place. Hokkaido Prefecture was almost certainly more inclined to grant permits for ceremonial harvest in the Monbetsu River because it already viewed the river's salmon as marginal—as outside of industrial hatchery and harvest systems—after it ceased hatchery releases there in 1994. While the logics for this particular management change are not clear, it was likely influenced by the more general conditions of the Hokkaido salmon industry at that time, which was suffering from a surge in salmon returns at the same time as price declines due to rising numbers of Chilean farmed fish (see the interlude and chapter 5). Yet while hatchery producers and industrial fishing abandoned the small river, the salmon did not. Enduring the river's concrete mouth and drainage ditch spawning beds, the former hatchery salmon kept on inhabiting the river, creating a population outside of the hatchery system that was large enough to support the Monbetsu Ainu group's salmon harvests but too small to draw industrial-scale attention or create conflicts with the nearby fishing cooperative.

In 2018, Hatakeyama-ekashi decided to further assert Ainu rights by fishing for the salmon for that year's *kamuy chep nomi* without securing a ceremonial harvest permit, maintaining that Ainu had never relinquished their right to salmon. Hatakeyama-ekashi and the other salmon ceremony participants arrived at the Monbetsu River to find prefectural police waiting for them. In a statement to the assembled group, Hatakeyama-ekashi spoke of Japanese colonial violences at the same time that he positioned Ainu claims in relation to transnational Indigenous movements: "I am one man among the world's Indigenous peoples. I have globally recognized rights to self-determination. That's why I'm doing this. The Japanese government is going against the flow of the rest of the world" (Kosaka 2019, 147).[30] The police arrested Hatakeyama-ekashi as he tried to lower his dugout canoe into the water, then subjected him to three days of hours-long interrogations before bringing criminal charges against him for harvesting salmon without prior permission (Indigenous Peoples Rights International 2020). In 2020, after Hatakeyama-ekashi suffered a stroke and was hospitalized, a district court suspended, but did not entirely dismiss, the charges, a move that some attribute to concerns on the part of Hokkaido Prefecture and the Japanese government about the negative publicity that the case could generate. The suspension of the indictment leaves the legality of Ainu fishing in limbo, as the court neither established nor rejected Ainu rights to harvest salmon (IWGIA 2021).

COMPARISONS TO COME

The ripples of Hatakeyama-ekashi's project, which mobilized a variety of alter-comparisons against those born from Japanese state colonialism, have been far from trivial. In August 2020, in solidarity with Hatakeyama-ekashi's efforts, the Raporo Ainu Nation in eastern Hokkaido sued Hokkaido Prefecture and the Japanese government to assert that their river-based salmon harvesting rights have never been extinguished by Japanese law (IWGIA 2021).[31] As a gesture of alliance, in early September that year, members of the Raporo Ainu brought salmon to the Monbetsu group so that they could hold a *kamuy chep nomi*, despite Hatakeyama-ekashi's hospitalization and their ongoing legal challenges (CEMiPoS 2020). While the trajectories of these solidarities and movements is uncertain, such new arrangements are likely to have multiple effects in the coming years.[32]

Hatakeyama-ekashi's initiatives have not only affected Ainu mobilizations; they have also impacted Monbetsu fish. It is important to remember that the rice, bread, and salmon of the Ainu man's classification in the earlier vignette are more than symbols. They index ecological assemblages. The comparisons with salmon that are part and parcel of Ainu rights movements at once depart from and affect more-than-human landscape arrangements. When Hatakeyama-ekashi began collaborating with various environmental NGOs, the Monbetsu River salmon were simply *sake*, or chum salmon. But through their joint work—including salmon surveys, water quality checks, and a salmon-focused workshop that brought Indigenous and environmental activism into closer conversation—the Monbetsu fish gradually became *wairudo sāmon*, a transliteration of the English term "wild salmon," and began stressing their non-hatchery origins. While the shift was discursive, it is not merely so. Because of their connections with Hatakeyama-ekashi and the Monbetsu Ainu, these salmon are likely to become different beings. Before the *kamuy chep nomi* and the Ainu rights movement, few people paid much attention to the river's salmon. Now, they are on the radar of several metropolitan environmental NGOs that seek to build networks between Hatakeyama-ekashi and biodiversity conservation initiatives, and their innovative propositions for new forms of salmon management explicitly build ecological sustainability outward from the practices of *kamuy chep* (Kamuycep Project Research Group 2021).

Exactly what this will mean for the fish is unclear, but it is possible to hazard a guess. Already, the pollution control agreement, which subjects the waste facility to extra surveillance, is helping to protect the river's water quality, increasing the odds that fish will survive there. Furthermore, if their rights movements are able to gain any traction, the Monbetsu Ainu group would like to take a more substantial role in watershed management, potentially altering forests and river habitats in other ways. If the number of salmon in the watershed grows in response to such changes, it would probably have substantial follow-on effects, as the carcasses from post-spawning salmon nourish organisms from stream insects to birds of prey to the brown bears, who are central parts of Ainu spirit worlds. Barring too many strays from nearby hatcheries, the river's former hatchery salmon are also likely to adapt to its specificities and develop unique place-based traits, their intergenerational futures taking a different path than they would have without Hatakeyama-ekashi's interventions. Bound in co-constitutive relations for hundreds of years, Ainu and salmon continue to recursively transform each other.

The relations of Ainu and salmon in Monbetsu show some of the challenges of living with the ongoing legacies of modernist (and statist) comparative practices. Fortunately, despite its concerted efforts, the Japanese state has not been able to completely control either the fish or Ainu-salmon relations. If everything had gone according to its plans, there would be no Ainu or free-spawning salmon in the Monbetsu region—only homogenized "Japanese citizens" and industrialized hatchery fish. But both are there. Persistence, though, has not been easy. The Monbetsu Ainu group and the Monbetsu salmon cannot opt out of the comparative structures that the state has used to render them marginal. Instead, they compare against them in creative and determined ways as they explore possibilities for remaking worlds rent apart by settler-colonial and industrial projects.

Coda

Embodied Comparisons beyond Japan

O N a late autumn day, I watch as two fisheries biologists in heavy-duty vinyl-coated rain gear collect decomposing salmon bodies along a small Columbia River tributary lined with alders. They haul the fish carcasses to a wooden worktable set up beneath the corrugated metal roof of a two-sided shed. With a sharp bread knife, one of the biologists saws vertically into the head of one of the salmon, about an inch behind its eyes. The fish's body is spotted with fungus, and it smells unmistakably of rotting flesh. The salmon is one of many who have returned to the creek to spawn, dying shortly after laying their eggs or releasing their sperm. One of the biologists cracks the head of the salmon over the edge of the table, peering into the brain cavity exposed by his cut. Swapping the knife for a pair of tweezers, he gently reaches into the lower part of the brain cavity and removes two soft sacs. Inside each is a small white stone: an otolith, or fish ear bone.[1] After wiping them on a paper towel, the biologist places them into small, carefully labeled vials that will be shipped to a university lab. He repeats the process again with another fish, and this time I ask if I can hold one of the bones, which reminds me of a sliver of broken seashell.

The tiny otolith that I cradle in my palm illustrates how comparisons literally matter. It shows how practices of comparative landscape-making find their way into the material bodies of fish. Composed of calcium carbonate and trace minerals deposited into a protein matrix, this otolith helped its fish to "hear." Although fish do not have eardrums, their otoliths work similarly to those in humans, turning sounds and spatial orientations into neural impulses as these small bones bump up against the hair cells inside cochlea. Unlike human otoliths, however, salmon ear bones continue to enlarge throughout the life of the fish. Otoliths are a fish's

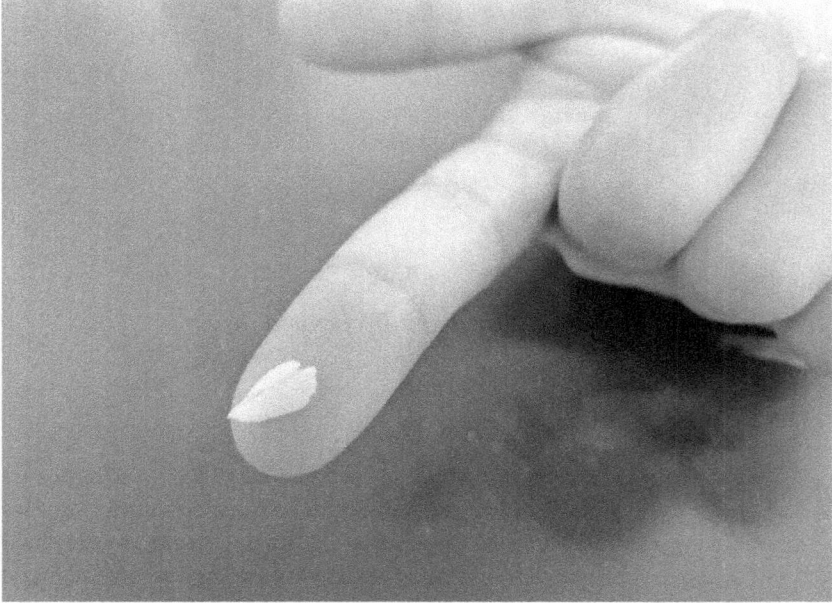

Salmon otolith. Photograph by George Whitman and Kimberly Evans. Used with the permission of the Center for Watershed Sciences at the University of California, Davis.

diary, accumulating like tree rings but at a faster rate. As a salmon grows, it lays down approximately one otolith band per day, and when examined in a laboratory, the width and chemical composition of these bands can provide a sense about different aspects of a fish's life. The bands change depending on its diet, migration routes, and levels of stress.

Forms of environmental management—including hatcheries, river alterations, and ocean fishing regimes—affect otolith deposition through the ways they alter the conditions of salmon, including what they eat, where they can hide, and the temperature of the waters in which they swim. Resulting changes in otolith patterns—like otoliths themselves—initially appear small. For example, a large number of salmon otoliths now display a "stress check," a dark heavy band deposited on the day they are released from a hatchery, testifying to the metabolic shock of moving from a tank to a river; stream-born fish, in contrast, have no such mark. Furthermore, some hatcheries intentionally create unique patterns of marks on the otoliths of their fish by varying the water temperature during their egg stage, essentially creating an internal barcode for the fish that makes them identifiable when otoliths are used in research projects. The otolith patterns of salmon—both hatchery and non-hatchery—have also shifted in other ways. In some regions, they have fewer freshwater and brackish-water bands in comparison to those of their ancestors, as the salmon spend less time in river and estuary habitats that have become increasingly developed and barren, with less food and fewer hiding spots for young fish. These altered patterns and the fish histories they index become visible through forms of otolith analysis that are themselves comparative; at the same time, they lead us back to the comparisons of river industrialization and hatchery production.

Practices of comparison are among the fragmentary stories that otoliths inscribe. Via the thoroughly comparative practices of landscape-making, the forces that social scientists often term political economy—such as industrial fishing, nationalistic claims to ocean resources, and land-based capitalist developments that degrade fish spawning grounds—shape the metabolic lives of fish and are thus calcified into these bones inside their heads. At the same time, comparisons also seep into salmon in other ways. While scientists have found otoliths especially useful for studying certain changes in salmon lives, there are additional effects of comparative landscape-making projects that are more clearly visible in fish body shapes and sizes, the timing of salmon returns, the location of spawning, the population numbers of different salmon groups, and genes and gene expression.

Without attention to comparisons, we simply cannot understand the bodies and lives of Hokkaido salmon or the watersheds in which they spawn. Many times over, comparisons have shaped Japanese salmon and their watersheds by creating new relations. Each chapter in this book has shown us how comparisons produce practices that remake salmon bodies, populations, and metapopulation structures within and beyond Hokkaido. Attuned to comparisons, we noticed how those between Hokkaido and the American West compelled the introduction of specific kinds of hatchery techniques and the development of a form of scientific fisheries management that led to particular production practices in Japan. We saw how Japanese desires to create a postwar economy comparable to that of the United States, as well as Japanese comparative assessments of Latin America, aided the formation of the Chilean farmed salmon industry, an industry that in turn has completely reconfigured Chile's ecologies. Back in Hokkaido, we observed how salmon populations were remade by the changes in global fish markets that comparisons between Japanese salmon and Chilean farmed fish engendered, and we traced how the island's fish have been shaped by the management practices of Japanese fishing industry professionals committed to being "modern" rather than either "traditional" or "out of date." Finally, we noted how comparisons that track through "wildness" and "indigeneity" have generated new conservation initiatives and fostered practices of river restoration. Because these kinds of comparative practices have caused such major changes in salmon morphology, genes, and population structures, noticing practices of comparison are an essential part of noticing Hokkaido salmon. By altering salmon worlds, comparisons shift the bodies and evolutionary trajectories of these fish. In the case of Hokkaido's salmon, these comparisons have had such strong effects that it seems appropriate to think of these fish as creatures of comparison.[2]

COMPARISON IN AND FROM HOKKAIDO

This book has attempted to highlight the role that comparative practices play in landscape transformation, in an attempt to cultivate a genre of multispecies political economy that follows the effects of industrial processes and landscape changes into the tissues of other-than-human species. To do so, it has looked at situated practices of comparing in and with Japan. Comparison, as a phenomenon, is in no way unique to Japan; on the contrary, it is a nearly ubiquitous act. Yet practices of comparisons take on very different geometries and textures within particular webs of relations. This book's aim has been to spark broad reflections about the role of comparisons in landscape-making by tracking the specific comparisons that have emerged with projects of making and contesting "modern Japan." For hundreds of years before the Japanese settlement of Hokkaido, Ainu management practices were coeval with the region's salmon. After 1869, however, its salmon were made *Japanese*, as they were pulled into new, explicit projects of state-making. Within this history, the evolutionary pressures on salmon become inseparable from nationalist modernization policies that were continually reaching out to places beyond Japan through acts of material comparison. The bodies of Japanese salmon, in the flesh, bring us into histories of comparative nation-making and landscape-making in an uneven world, serving as a reminder that geopolitics matter, literally, to the bones and tissues of other-than-human beings. Although it is rarely phrased in such ways, Hokkaido salmon genes are fundamentally shaped by nineteenth-century Japanese fears of Euro-American colonization and the colonization of Hokkaido that they enacted in response, as well as by the twentieth-century politico-economic dynamics of post–World War II high-seas salmon fisheries, Japanese development aid and supply chain management, and twenty-first-century transnational conversations around environmental conservation and Indigenous rights.

The terms through which geopolitical dynamics are expressed can often be problematic and ahistorical, with erroneous elisions between ethnicity, culture, and nation-state practices. In everyday encounters in Japan, comparisons between Japan and the West are frequent and frequently stereotypical. But this seemingly binary civilizational comparison is not as singular, generic, or categorically rigid as it might initially seem. While it is often invoked in sweeping terms, it is also iteratively brought into being together with comparisons among specific places. When we look at comparisons between Japan and the West in the flux of everyday life, we see that they at once emerge out of and are constantly interrupted by complex and multidirectional webs of comparative practices that draw in Hokkaido, the Columbia River's salmon canneries, the tastes of English and French foreign service members, a river in southern Chile, and the many more sites that have appeared in this book. A Hokkaido salmon, then, is Japanese in the sense that it has been shaped by projects that intentionally sought to build a modern Japan, but that Japanese-ness is neither innate nor wholly located in Japan; instead, it is emergent out of transnational comparisons that have historical patterns but that are also contingent, creative, and heterodox.

BEYOND SALMON AND BEYOND JAPAN

Such phenomena are relevant beyond salmon and beyond Japan in many ways.[3] Related comparative landscape-making dynamics are at play for many non-Western countries caught in the complex comparisons of modernity-making, as "specters of comparison" (to use Benedict Anderson's [1998] phrase) are common to projects undertaken in the name of progress. Because the vast majority of development projects routinely swap models and envision futures through the presents of other places, similar strategies for considering comparison are likely relevant for exploring more-than-human relations in other parts of Asia, Africa, and Latin America. But what about areas *within* Euro-America, within that which one might call the West? Are they, too, remade by comparisons, even if those comparisons are sometimes hidden or harder to see? If so, how?

I was at first startled by all of the overt comparisons I encountered in Hokkaido salmon management because they were so rare in the salmon worlds that I knew from living and working in the US Pacific Northwest. My own comparison between the Columbia River and Hokkaido compelled my attention to comparison in Japan. Yet it also made me curious about the seemly *non*-comparative nature of Columbia River salmon worlds. Were they really as un-comparative as I initially thought? Or was I just failing to notice the comparisons within them? The otoliths mentioned in the first pages of this coda were extracted from Columbia River fish. What comparisons, if any, had shaped their formation, along with the bodies and lives of the fish from which they had been extracted?

When nineteenth-century Japanese colonial officials compared Hokkaido and its salmon to those of the Columbia River, the comparisons were indeed largely one-way. In searching several of the important repositories for Columbia River fishing-related archives, I have found no evidence that American officials expressed any interest in learning about Hokkaido's fisheries at that time. Neither have I found any US notes about the visits that Japanese officials and their emissaries made to Oregon and Washington as they developed Hokkaido's salmon industry. Their curiosities were not the same. While Oriental art and lacquerware captured the imaginations of urban Euro-American elites who were tickled by oddities of those they framed as exotic, the nineteenth- and early twentieth-century development projects of Japanese officials do not seem to have piqued the interests of the Americans seeking to develop the Columbia River region. Why was a visit from a Japanese official not seen as a noteworthy opportunity to learn about Japanese fisheries? And why, after all, were there no American missions to learn about fisheries in Hokkaido until the 1980s, when the state of Alaska wanted to know how Hokkaido's fish hatcheries had come to so dramatically outperform theirs?

Perhaps the answer lies in the ways that the development of the American West has been fundamentally entangled in assertions of national noncomparability. National exceptionalism has a storied place in American thought; the United States was founded on claims of divine guidance and radical experimentation as the young country tried to position itself as a break with Europe, as different from its established ways.[4] As (white) Americans have imagined their nation as one of incomparable greatness, US popular narratives have tended to suppress rather than celebrate the transnational comparisons that have been integral to the formation and development of the United States. But while they are typically absent from historical accounts, concrete comparative projects have indeed played key roles in shaping western American landscapes. For example, at the same time that the Hokkaido Colonization Commission was importing new species of plants and animals from the United States, the US Department of

Agriculture was sending its own plant collecting expeditions to Asia (Chacko 2018). But in contrast to Hokkaido, where the histories of crop-plant introductions are widely known (albeit problematically framed within celebratory colonial narratives), the origins of crop plants naturalized to the United States have largely been erased.

Traces of comparisons, however, clearly remain, not only in landscapes but also in some archives. Despite the paucity of records in the Columbia River–Japan case, some nineteenth-century Americans did indeed record comparative modes of envisioning the American West. For example, George Perkins Marsh, an American diplomat who had spent time in the Ottoman Empire, saw the drylands of the American West in comparison with Arabian deserts. For Marsh, that comparison led him to strongly support US military efforts to deploy camels in an effort to remake American desert worlds. In an 1854 address to the Smithsonian, Marsh elaborated his comparative thoughts:

> The habits of the Indians much resemble those of the nomadic Arabs and the introduction of the camel among them would modify their modes of life as much as the use of the horse has done. For a time, indeed, possession of this animal would only increase their powers of mischief; but it might in the long run provide the means of raising them to that state of semi-civilized life of which alone their native wastes seem susceptible. Products of the camel, with wool, skin and flesh, would prove of inestimable value to these tribes, which otherwise are likely to perish with the buffalo and other large game animals; and the profit of transportation across our inland desert might have the same effect in reclaiming these barbarians which it has had upon the Arabs of the Siniatic peninsula. (Marsh 1855, 120)

Marsh's settler-colonial comparative practices at once resonate with those of late nineteenth-century Hokkaido officials and differ from them. Marsh, like Hokkaido officials, was comparing in the name of colonial practice—of violently destroying Indigenous lifeways and fostering economic development. But he was able to compare with a sense of surety that Hokkaido officials did not have. Marsh did not worry if the overall development and modernity of the American West would measure up to that of the Arab world. His comparisons were marked by the confidence of comparing from a transnationally recognized position of power. Beyond Euro-America, comparisons are more anxious as they are judged not only within the frame of one's own nation-making but also by other more powerful nations. Such comparative unease does not in any way excuse or mitigate the violences of Hokkaido colonialism. Instead, attention to it is a tool for analyzing the specific ways its violences unfold.

Perhaps counterintuitively, attention to comparisons may be as important for understanding Columbia River salmon and their management as it is for Hokkaido salmon and theirs. For Japan, the challenge of capitalist modernity has been one of becoming *comparable*. In contrast, within US narratives, it has been framed as one of becoming *incomparably* great. Yet such assertions of incomparability are nonetheless simultaneously built out of, justified through, and challenged by everyday comparisons that reach across space and time. The Columbia River salmon canning industry was, of course, shaped by the rural men in coastal Scandinavia and Finland who heard about the comparatively more lucrative fisheries in the Columbia River and decided to emigrate. But it was also constructed from settler-colonial comparisons that justified usurpation of Indian fisheries, as well as from other racialized comparisons that justified the recruitment and second-class treatment of its Chinese contract laborers. Furthermore, as in Hokkaido, there were also counter-comparisons that challenged salmon industry practices, not only its racialized and gendered economies but also its environmental effects. In the case of US Pacific Northwest salmon, a 1921 article in a popular regional magazine, unconvinced by the alleged promise of hatcheries, was already comparing US fish to "the bison, the passenger pigeon, and the great auk" as other parts of the United States and North Atlantic worlds were seen as harbingers of the problems that industrial salmon fisheries were likely to create (*Sunset Magazine* 1921).

How is one to see the traces of comparisons that are often overlooked in nations that are still reluctant to be haunted? Other anthropologists and historians have begun to probe this question by tracing the practices of comparison within which US-based projects are iteratively made, alongside their elisions.[5] Following their lead, if I were to do research in the Columbia River now, after my encounters with Japanese salmon, I would approach the question of how salmon are done there with a different sensibility. I would pay far more attention to the erasure of transnational connections, and I would not take the largely self-referential quality of doing Columbia River salmon at face value. Instead, I would try to notice the practices of exclusion through which the unmarked categories and ostensibly un-comparative worlds in the Columbia River are made, while querying how the US Pacific Northwest has been able to become such a seemingly insular salmon world. Part of this practice would also be to listen more closely for specters of comparison, asking how the salmon worlds of other places haunt those of the Columbia River basin. Although they are not made overt in everyday practices of doing salmon, within American hatcheries, restoration projects, and laboratories there are hints of hauntings that more attuned eyes and ears might catch and query: American scientists who dismiss Japanese work as irrelevant, Pacific Northwest tables filled with salmon from Chile, and hatchery salmon feed that contains protein from Peruvian anchovies.

Regardless of whether a context resembles Hokkaido or the Columbia River, attention to comparisons serves as an important hinge in the ways that it better enables social scientists to integrate research on nation-making and transnational encounters with that of multispecies and more-than-human scholarship. Comparisons, along with assertions of incomparability, warrant more attention as powerful but often overlooked landscape-making forces, ones that fundamentally transform the lives and bodies of other species. It is not enough to examine comparisons within human social registers. We must also follow them into more-than-human worlds.

NOTES

1 The vast majority of the salmon in Hokkaido—and thus in this book—are *Oncorhynchus keta*, commonly referred to as *shirozake* in Japanese, *kamuycep* in Ainu, and chum salmon in English. There are smaller numbers of commercially harvested pink salmon (*Oncorhynchus gorbuscha*) in Hokkaido, but these are referred to as trout (*karafuto masu*) in Japanese. Hokkaido is home to still smaller numbers of cherry salmon (*Oncorhynchus masou*), also considered trout in Japanese (*sakura masu*), but these are not a commercially significant fish.

2 This approximation is based on 2008–18 data from the Japan Ministry of Agriculture, Forestry, and Fisheries 2020.

3 This book uses Japanese honorifics, primarily *-san* (roughly the equivalent of Ms. or Mr.), for Japanese-language speakers, as these were the name conventions within used ethnographic contexts. Except where otherwise noted, names are pseudonyms. At her request, this is Miyoshi-san's real name.

4 I use the terms *multispecies* and *more-than-human* more or less interchangeably. While there are concerns that *multispecies* positions scientific ways of knowing as an unexamined norm, I use it alongside *more-than-human* in the context of this book, as nearly everyone I interviewed during its research uses the concept of species (even as they also draw on ways of knowing that do not track through the scientific). The term *worlds* is a widespread albeit imprecise concept in anthropology that does not fully align with its use in philosophy. As used here, *worlds* are material and relational; they are not static but rather continually brought into being within practices.

5 See Kolbert (2014) on the Sixth Extinction; see also Lewis and Maslin (2018), Lorimer (2017), and Swanson, Bubandt, and Tsing (2015) for different overviews of the Anthropocene and its social lives.

6 This shift toward the study of entities and beings beyond the human has become a wide-ranging movement across the humanities. While this book can be read as part of this general movement in the humanities, sometimes called "the material turn," it focuses less on material agency and more on material historicity.

7 The lack of ecology in political ecology has been a long-standing topic of conversation (P. Walker 2005).

8 This paragraph is indebted to the thinking of Anna Tsing and the Aarhus University Research on the Anthropocene project.

9 This book is also substantially influenced by other modes of environmental and animal history, such as Ritvo (1987) and Anderson (2004).

10 See Gluck (2011) for a powerful explanation of modernity as a historical process, not mere trope, and for discussion of the forms of "improvisational modernity" that arose in Japan.

11 For an overview of Hokkaido's settler-colonial history and the need to challenge its common narratives, see Grunow et al. (2019).

12 In a classification of nations published by a Japanese government body in 1869, "Russia was not put into the highest category of 'civilized countries' (*bunmei no kuni*) together with England, France, the Netherlands, and the United States (later joined by Austria, Prussia, Denmark, and Sweden). Russia, along with Italy, Spain, Portugal, and the countries of Latin America, was placed in the second category, 'enlightened countries' (*kaika no kuni*). From there on, the list descended as follows: China, India, Turkey, Persia, and the African nations north of the Sahara were classified as 'semi-enlightened countries' (*hankai no kuni*), while the nomadic tribes in Siberia, Central Asia, Arabia, and Africa were classified as 'countries of uncivilized manners and customs' (*izoku no kuni*). Last came the 'barbarians' (*yaban*): the American Indians and the natives of Africa and Australia" (Togawa 1995). As a consequence of this categorization, the Japanese government sent few officials and students to Russia, and only one person with Russian travel experience was selected to serve in an important government position (215).

13 The primary fieldwork for this book was conducted from August 2009 to December 2010, with short follow-up trips in 2011 and 2015. Preliminary research also occurred in 2006–8.

14 This paragraph draws on the work of Liu (1995), Stanlaw (1992, 2004), and Hogan (2003), who specifically studied how people in Hokkaido incorporate English words. These scholars reject descriptions of katakana as "borrowing" or "loanwords" in any simplistic sense, stressing instead the creative and inventive making of katakana terms.

15 The genus *Oncorhynchus* developed in the early Miocene (15–20 Ma), compared to approximately 300,000 BP (before present) for *Homo sapiens*. Even if one wants to define the emergence of salmon through

species rather than genus, they are still far older than people. According to fossil evidence, present Pacific salmon species all evolved prior to 6 million years ago (Waples, Pess, and Beechie 2008).

16 This section is indebted to conversations with Frida Hastrup and Nathalia Brichet.

1. SITUATING COMPARISONS

1 By 2020, however, the number of Honshu salmon had declined more sharply than Hokkaido fish, so there are now closer to ten times as many salmon in Hokkaido as in Honshu. Statistics from Hokkaido National Fisheries Research Institute (2020).

2 This point builds on that of other scholars reconsidering comparison in light of Viveiros de Castro's work, including Jensen et al. (2011), Gad and Jensen (2016), Jensen and Morita (2017), and Mohácsi and Morita (2013).

3 Ilocano refers to a Filipino ethnolinguistic group with ties to the Ilocos region, the northwestern part of the island of Luzon, which was subject to Spanish colonization efforts from the sixteenth century onward.

4 The depictions of these countries by the Japanese participants did not fully grapple with their actual practices, which include oil extraction and ongoing battles over Sami rights.

2. LANDSCAPES, BY COMPARISON

1 Translation roughly based on Petersen (2007), but slightly modified by the author.

2 Translation from Petersen (2007).

3 What places counted as Ezo also varied according to the historical moment; while Ezo generally included most of the island known today as Hokkaido, as well as those known as Sakhalin and the Kurils, it was indeterminate, often expanding and contracting depending on who drew the map (Edmonds 1985; Morris-Suzuki 1998b).

4 The transformation of Ezo into Hokkaido did not happen overnight. For several decades, both names were used, often with confusion. For example, a 1902 missionary report indicates that Ezo was used to refer to the main island, while Hokkaido referred more generally to all of Japan's newly claimed northern lands (Batchelor 1902).

5 While Hokkaido eventually came to denote a fixed district that encompasses the main northern island, Japan's northern boundaries did not become static (Morris-Suzuki 1998b). Disputes with Russia over the ownership of the southern Kurils continue, and although the Japanese

government is not actively pursuing claims to its former colonial lands in Southern Sakhalin, it continues to assert that the question of sovereignty in this area has not been officially settled.

6 As Lu (2019, especially chapter 1) describes, Japanese officials—drawing on Malthusian logics—also compared Hokkaido immigration to the founding of the United States, invoking the story of the *Mayflower* and the Puritans.

7 For more on Nitobe and his time at Sapporo Agricultural Collage, see Dudden 2019.

8 During the Meiji era, people took note of such differences as they tried to make sense of Hokkaido. Thomas Blakiston, a Briton who lived in Hakodate from 1861 to 1884, concluded, based on his natural history observations, that "Yezo and more northern islands are not Japan, but, zoologically speaking, portions of northeastern Asia, from which Japan proper is cut off by a decided line of demarcation in the Strait of Tsugaru" (Blakiston 1883, in Cortazzi 2000, 154).

9 In 1870, the Japanese government recommended the following countries as models for exchange students interested in specific fields, providing a sense of the diverse comparisons that Japanese government officials made: Britain (machinery, geology and mining, steelmaking, architecture, shipbuilding, cattle farming, commerce, poor-relief); France (zoology and botany, astronomy, mathematics, physics, chemistry, architecture, law, international relations, promotion of public welfare); Germany (physics, astronomy, geology and mineralogy, chemistry, zoology and botany, medicine, pharmacology, educational system, political science, economics); Holland (irrigation, architecture, shipbuilding, political science, economics, poor-relief); and the United States (industrial law, agriculture, cattle farming, mining, communications, commercial law) (Nakayama 1989, 34).

10 The Meiji government (and the Tokugawa Shogunate in its final years) sent government officials and students abroad, with the number of people dispatched varying by year, ranging from tens to a few hundred annually (Hara 1977; Inoue 2008).

11 Nitobe, who studied for three years at Johns Hopkins University, frequently wrote about Japan for American and other English-speaking audiences and also authored texts in German.

12 What Capron refers to as native Japanese horses are those primarily descended from continental Asian populations and specifically bred for millennia in Japan (International Museum of the Horse n.d.) Cattle have been similarly reared in Japan since around 200 CE, with distinct island breeds emerging from continental Asian populations (Mannen et al. 1998).

13 Fruit trees and seeds were also distributed to other parts of the Japanese isles (Walker 2004, 256). Transfers of plants and animals were part of a

widespread nineteenth-century interest in "acclimatization," or the introduction of species to new locales. See Dunlap (1997) and Lever (1992) for general information on acclimatization.

14 While the differences in meanings attached to mammal meat, particularly beef, in Europe and Japan before the late nineteenth century are clear, the rates of actual meat consumption are not. It appears that people in Japan may have eaten a fairly substantial volume of hunted meat at various time periods (Krämer 2008).

15 The director of the Tokyo Naval Hospital and the Head of the Bureau of Medical Affairs of the Navy, who beginning in the mid-1880s encouraged military beef eating, had studied in London for five years (Cwiertka 2002, 9–10).

16 While the Appropriations Act of 1851 authorized the creation of the Indian reservations, later nineteenth-century policies often emphasized land privatization via allotment over removal to reservations. The Dawes General Allotment Act of 1887 is one example of this shift. In the twentieth century, policies again vacillated between termination and recognition of tribal rights. For a description of Clark and other American advisors' views on Ainu, race, and settler-colonial practice, see Hennessey 2020.

17 Capron's memoirs indicate that while Ainu people reminded him of American Indians, he saw them as more amenable to civilization. Capron praised what he saw as signs of Ainu adaptation to agricultural settlement: "Vegetables and fruits now supplement the meager diet of fish and sea weed of the native Aino, and his simple expression that 'potatoes go so good with fish' speaks volumes of encouragement to the Japanese promoters of this Commission" (Capron 1884, 305). What Capron likely did not realize was that Ainu have an agricultural history stretching back to at least the ninth century, including grains, vegetables, and indeed, potatoes (Crawford and Yoshizaki 1987). Potatoes were part of Ainu agriculture prior to the Meiji period, perhaps from their introduction to Hokkaido in 1706 and certainly from the early nineteenth century (Hosaka 1993).

18 This sentence focuses on government-run Ainu elementary schools as described in Tanabe 2019. The Anglican Church Missionary Society (CMS) also ran Ainu day schools from 1888 to 1906, along with the Hakodate Ainu Training School (1893–1905), a boarding facility. In contrast to the government-run schools, these institutions included Ainu language coursework (Tanabe 2019). By 1910, more than 90 percent of Ainu children were attending school, with roughly one-third of school-age Ainu children at government Ainu schools and the remaining two-thirds at other institutions, including CMS Ainu schools (Ogawa 1997, in Tanabe 2019).

19 Hokkaido remained under direct control of the central government until after World War II, when it became a regularized prefecture.

20 Military drill was a required part of the curriculum, with the goal of cultivating bodies in addition to minds. The Hokkaido colonial government, fearful of Russian incursions, was also interested in ensuring that its population was ready for military mobilization. Military training had been included in the original Morrill Act, as it was passed by Congress shortly after the start of the US Civil War (Abrams 1989).

21 For more on the relations between "pioneer spirit" and Christianity in Hokkaido, see Shirai 2010.

22 See exhibits about food items at the Sapporo Clock Tower Museum.

23 The first formal institutes of higher education were not established until the mid-nineteenth century. The oldest institutions are Keio University (1858) and Tokyo University (1867). Thus, when SAC was established, higher education in Japan was still in its infancy.

24 The article from which this statistic is taken raises important questions about the role of Chinese merchants in Hokkaido, alongside American influences.

25 The phrase *ethnic Japanese* is used to identify Japanese people from Japan's southern islands, vis-à-vis Ainu peoples, who were made Japanese citizens and often self-identify as Japanese as well as Ainu.

26 Although the Columbia River sparked the salmon boom, its production was quickly eclipsed by that of Alaska. By 1901, Alaskan canneries were producing nearly ten times as many cases of fish, albeit at a lower quality and price (Martin and Tetlow 2011, 19).

27 In 1877, in addition to salmon, the facility also produced canned venison (9,358 cans), canned oysters (3,226 cans), and canned beef (Treat 1878).

28 Canned food products were slow to catch on in Japan and never reached the popularity that they did in European countries (Cwiertka 2006, 61).

29 The differences in taste and texture that Euro-Americans noticed between Japanese and American canned salmon products can be explained in a variety of ways. The regions used different species of salmon with markedly different flesh consistencies and oil content. In addition, the use of different kinds of salt and different canning technologies also likely produced substantially different tastes. The "made in Japan" labels attached to such products may have also influenced Euro-American taste testers and may have led Euro-Americans to interpret differences between American and Japanese salmon products as inferiorities on the part of the Japanese goods.

30 Clark 1877b, page 11 in the digital archive numbering, sheet 6 as hand-numbered by author.

31 This book refers to this island by its internationally recognized name of Sakhalin. However, the Japanese speakers with whom I interacted often

used Karafuto, its name under Japanese rule. Similar tensions exist for the names of the Kuril Islands, as Japan continues to dispute Russian claims to the four southernmost islands, referring to them as the Northern Territories. These regions also have Ainu and other Indigenous language names.

32 This information about the official's encounters with hatchery technologies in Vienna comes from a summary of an exhibit at the Saitama Prefectural River Museum (Saitama Kenritsu Hakubutsukan 1998), as well as from Wada (1994). Although hailed as a model at the Vienna exhibition, the Australian attempts to introduce salmon ultimately failed to produce self-sustaining runs of these fish (Lien 2005).

33 Information about Ito comes from displays and conversations with staff at the Chitose Sake no Furusatokan (Chitose salmon aquarium) in Chitose, Hokkaido, as well as from Ichiryūkai (1987).

34 The number of places Ito visited and the diversity of fisheries he observed was immense. See his itinerary, reprinted in Ichiryūkai (1987). See also Ito's original report (1890).

35 For more on the Columbia River fish wheels that Ito saw during his trip, see Seufert (1980).

3. OF DREAMS AND COMPARISON

1 Other well-known SAC graduates in colonial governance include Kawakami Takiya, who became a botanist with the Taiwanese colonial administration, and Tōgō Minoru, a high-ranking bureaucrat in colonial Taiwan and noted proponent of Japanese racial supremacy. The allure and promotion of Hokkaido models also attracted interest from non-Japanese. In a 1905 document, Chinese officials explicitly advocated the opening of Chinese experimental farms based on those in and around Sapporo (Lawson 2015, 52).

2 Japanese fisheries managers built the first hatcheries in what is now Russia in the 1920s (Nash 2011, 88).

3 JICA is roughly the equivalent of the US Agency for International Development (USAID) or Germany's Gesellschaft für Internationale Zusammenarbeit (GIZ; the German corporation for international cooperation).

4 I base my description of the JICA-Chile project primarily on interviews with Nagasawa-san, other Hokkaido fisheries scientists who traveled to Chile, Chilean participants in the JICA project, and JICA officials. My understanding has also been enhanced by Hosono (2010).

5 Japanese involvement in Chile has focused more on resource acquisition than on colonial settlement. In contrast to Peru and Brazil, the Japanese government did not send emigrants to Chile, nor did the Chilean government solicit Japanese workers.

6 The success of these early efforts is debated. Some sources say that they did not create lasting runs of fish, while others hail this moment as the beginning of Chilean trout populations. Academic sources (e.g., Urrutia 2007) tend to be skeptical of nineteenth-century successes.

7 The Japanese members of the JICA-Chile salmon project all cited this historical event as one of the reasons that Japan and Chile have good relations.

8 An exact accounting of how the industry came to be eludes even those who try to study it directly. The author of one article, which set out to identify the main actors and factors that brought about the Chilean salmon industry, ultimately concluded that due to the large number of intertwined people—government groups, private businesses, and individuals—the precise origins of the sector could not be determined (Urrutia 2007, 463). The best the author could do, he said, was to allude to the "grand diversity" and "heterogeneity" out of which the industry was born (463).

9 Quotes are the author's translations from interviews and conversations that took place primarily in Japanese but with some use of English and occasional Spanish.

10 This trip was sponsored by the Japan Overseas Technical Cooperation Agency, which was the precursor to what is now the Japan International Cooperation Agency (JICA).

11 In their ocean life phase, salmon eat krill, along with squid and smaller fish, such as herring, anchovy, and sand lance.

12 The date when these efforts began is unclear, but in 1982, they succeeded in harvesting the first eggs and milt (semen and seminal fluid) from salmon reared to reproductive adulthood in Chile (Hosono 2010, 46).

13 The reasons that efforts to naturalize chum salmon in Chile were not successful remains unclear from a biological perspective. However, salmonid species that transplant easily are the exception rather than the norm. Within the *Oncorhynchus* genus, only Chinook salmon and rainbow trout, out of the twelve currently recognized species, have successfully established self-reproducing populations in new places on a substantial scale. See Rossi et al. 2012.

4. THE SUCCESS OF FAILED COMPARISONS

1 During the fieldwork in Chile on which this chapter is based, interviews with Japanese traders were conducted by the author in Japanese, while interviewers with Chileans in the salmon industry were conducted either in Spanish (with an interpreter) or in English.

2 After the mid-1980s and the end of the JICA project, Norwegian companies began to have an increasing influence on the Chilean salmon
industry (Katz 2006). In 1987, via Norwegian interactions, Chileans
began to rear Atlantic salmon in addition to Pacific salmon (Phyne and
Mansilla 2003, 112).

3 *Hochare* is the Japanese word for a fish who has already spawned and
who is either approaching death or has recently died. In Aros's words,
the flesh of a *hochare* "has no color and no taste and it disintegrates."
However, the low oil content of *hochare* makes them valuable to Ainu
people, as they are easier to preserve via drying.

4 Although tinkering with salmon color has a longer history, the SalmoFan
is a trademarked product that became widely popular in 2003, when
Hoffmann-LaRoche, a company that manufactured salmon-feed supplements, included the fans for free with all orders. The SalmoFan is now
owned and produced by DSM Nutritionals (Cha 2004; DSM Animal
Nutrition and Health n.d.).

5 The species most commonly used in fish meal in Chile are anchovy and
horse mackerel, while in Norway, they are capelin, herring, and blue
whiting (Miles and Chapman 2006).

6 Estimated using historical exchange rates from FRED (2021).

7 *Furikake* are fish flakes often sprinkled atop rice.

8 In 2011, approximately 39 percent of Chile's salmon exports went to
Japan, 24 percent to the United States, 10 percent to Brazil, and
4 percent to Europe (Esposito 2011). By 2018, the United States constituted 27 percent and Japan 23 percent of Chilean salmon exports, but
Japan remains significant; in the same year, Chilean salmon farms
increased their production of coho salmon (185,000 metric tons that
year) (Salmon Chile n.d.). Because Japan is the almost exclusive market
for farmed coho, this shows a continuing interest in catering to Japanese
consumers.

9 Japanese companies continue to dabble in Chilean salmon farm ownership. In 2011, after the Fukushima nuclear disaster and disruptions to
Japanese fisheries, the Mitsubishi conglomerate purchased a Chilean
salmon farm (*Nihon Keizai Shinbun* 2011) and subsequently expanded
holdings in the region (White 2016).

10 Yamada-san sometimes criticized farmed salmon in general but often
made specific reference to Chilean-produced fish.

11 Producers contest descriptions of farmed salmon as dyed or artificially
colored, as their flesh color is controlled through levels of astaxanthin
in their feed. While the astaxanthin used in aquaculture is primarily
synthetically derived, it is the same compound that produces the pink
hue in wild fish when they ingest it via the bodies of krill and shrimp.

12 As a 2019 article in an industry e-magazine discusses, Chilean salmon farms' use of antibiotics is decreasing but remains high (Evans 2019). See also Arroyo 2017.

13 For more on salmon farming labor issues and health impacts, see Aguayo (2008) and Latta and Aguayo (2012). These articles correspond to what I heard during my own much shorter visit to this region.

INTERLUDE

1 In Alaska, for example, between 1984 and 2002, "real (inflation-adjusted) ex-vessel prices for most . . . species had fallen to about one-third of average prices during the 1980s" (Knapp 2007, 240–41). The salmon market glut affected all species but was particularly difficult for chum salmon, which consistently garner lower prices than species such as sockeye and Chinook. For the effects of imported farmed salmon on Japanese markets, see Shimizu (2005).

2 The notion that ties of transnational trade can remake more-than-human worlds is far from novel in the social sciences. Scholars have developed a wide range of concepts to highlight the ecological consequences of carving the planet into zones of production and consumption. For example, Immanuel Wallerstein's "core-periphery" relations (2004) have helped us understand how the extraction of raw materials from colonial regions has fueled the concentration of wealth in the metropolises of the Global North, while such concepts as the "ecological footprints" have highlighted the outsized marks that urban areas leave on their surrounding rural landscapes (Rees 1992). Yet attention to the effects of the Chile-Japan salmon trade on Hokkaido's ecologies pushes us to consider different geographies than those featured in most of such research.

3 For examples, see Freidberg (2004) on European vegetable imports from Africa, Mintz (1985) on sugar, and Pomeranz and Topik's (2014) short essays on a variety of commodity-chain histories.

4 Ishikawa and Ishikawa (2013) have made a similar move, showing how the transnational wood products trade has altered Japanese forest ecologies by reducing domestic timber harvests.

5 Hokkaido salmon populations continue to fluctuate. When the majority of my field research took place between 2008 and 2011, Hokkaido salmon returns ranged from about thirty-nine to forty-eight million fish. Those numbers had fallen substantially by 2018–20, when they ranged from eighteen to twenty-three million, a level to which they had not fallen since the early 1980s (Japan Ministry of Agriculture, Forestry and Fisheries 2020). These declines are discussed in chapters 5 and 6.

6 See, for example, Tsing (2005), as well as the citations in note 3 of this interlude.

1 In this chapter, I use the term *fisherman* when discussing ideas and practices that are seen as being male-specific by people in the salmon fishing cooperatives where I worked. For example, women are generally not allowed on salmon fishing boats in this region, so the emptying of nets is gendered male. However, women do own shares of salmon fisheries and participate in fish sorting and other dock work. Thus, when I refer to more general aspects of fish cooperative work, I use gender-neutral terms such as *fisherpeople* and *fishers*.

2 While the degree of self-management in salmon fisheries in Japan is very high in comparison to those in the United States and Canada, it is not in itself a unique arrangement. For an overview of self-governance and comanagement, see Townsend et al. 2008.

3 This term carries connotations of being "behind the times."

4 In the postwar era, American occupation officials encouraged such interpretations. In their reports, they described Japan's fisheries as something "handed down from the feudal era" and thus in need of modernization (Hutchinson 1951, 174). The United States played a significant role in postwar fisheries policies and cooperative organizational structures, even directing radio announcers to produce a series of broadcasts on how to enact properly democratic fishing cooperatives (GHQ/SCAP 1950).

5 The reading skills required for the two publications are also very different. For example, my eleven-year-old Japanese friend could already read the *Nikkan* but could not yet make much sense of the *Nikkei*.

6 For a description of this 1948–50 Japanese fisheries reform from a biased but historically interesting American perspective, see Seidensticker (1951). For more scholarly analyses, see Yamamoto (1995) and Makino and Matsuda (2005). Under this American occupation policy, previous fisheries ownership structures were replaced by "democratic" fisheries cooperatives with owner-fishers. This process paralleled a similar agricultural land reform, which distributed land rights to previously tenant farmers (Kawagoe 1999). In December 2018, Japan enacted a new fisheries law reform, the first in seventy years, with implications that are not yet fully clear.

7 Salmon fishing rights are hereditary in most contexts, but each cooperative independently decides what kinds of inheritance patterns are acceptable. Until the last decade or so, rights were typically passed from father to firstborn son, but inheritance rules have since become more flexible. In Kitahama, for example, widows, sons-in-law, grandsons, nephews, and daughters also have inherited rights. There have also been several cases of fishers who gained their rights through "adult adoption,"

a practice in which an adult becomes the legal child of an older person, taking that person's last name, caring for that person, and then inheriting his or her fishing rights.

8 While pink salmon are a minor part of commercial catches dominated by chum, other trout species do not play a substantial role in Hokkaido's commercial fisheries.

9 Although such set-nets were common in US West Coast salmon fisheries in the late nineteenth and early twentieth centuries, they were banned for commercial use in Oregon and Washington in the 1930s and in Alaska in 1959. Yet in 2021, Washington State re-legalized salmon traps under some conditions as they are increasingly viewed as a sustainable fishing method (Wild Fish Conservancy Northwest 2021). See Swanson (2019).

10 Yet in an immediate postwar moment characterized by food shortages and general instability, salmon set-net rights nonetheless seemed appealing enough to Kitahama residents that hundreds of people wanted them.

11 On top of shares, the board members have also created a bonus system that gives small extra rewards to the members who serve as dockworkers and boat crew for the boat with the year's largest catch because they end up with the most work of unloading and sorting fish.

12 Every year, the group's board members go on a comparative study tour (*kenshū*) to enhance their understanding of global fisheries. When their harvests are good, they travel internationally, and when I was there, they were debating if they should travel to Australia or Vietnam.

13 Kitahama sells its fish to a variety of wholesale traders and companies through daily auctions. Fish auctions are common in Japan, most famously the tuna actions of Tsukiji Fish Market, described in Bestor (2004).

6. WHEN COMPARISONS ENCOUNTER CONCRETE

1 See Hébert 2010 and 2015 on changes in the Alaska salmon industry.

2 For a history of fisheries science ideas with a focus on salmon, see Bottom 1997.

3 Segawa 2007 and personal communication.

4 The facility was eventually forced to clean up its act by installing a settling pond and a water treatment process.

5 See chapter 1 in McCormack ([1996] 2016) for a broader discussion of Japan's "construction state," as well as Kerr 2001 for a description of the role of concrete in Japanese modernization efforts.

6 About twenty people participated in this event, but the society has about 150 active members.

7 North American salmonid species, such as rainbow trout, and European fish, such as brown trout, were introduced to Hokkaido in the late nineteenth and early twentieth centuries (Hasegawa 2020).

8 The scientific names are *Oncorhynchus nerka* (sockeye), *Oncorhynchus kisutch* (coho), and *Oncorhynchus tshawytscha* (Chinook).

9 Kitada (2014) estimates that the majority of hatchery chum salmon (87 percent) are produced by private hatcheries, with the remainder (13 percent) by national hatcheries.

10 These practices were not unique to Hokkaido. See Taylor (1999) for a description of cross-river egg transfers in the United States. In the Columbia River, hatchery workers, worried that they might not fill their quotas of eggs if they waited until late in the season, also used the earliest returning fish as brood stock. As a result, the genes of early returning fish are also overrepresented there, and over the course of several decades, the timing of hatchery salmon runs has crept earlier (Quinn et al. 2002).

11 Alaska, with large-scale chum and pink salmon runs, has a somewhat different history. In the 1970s, the state of Alaska took notice of Japan's hatchery success. Until that decade, Alaska, one of the world's largest salmon producing regions, relied on stream-based salmon reproduction, constructing only a handful of hatcheries in the state's southern panhandle. But in the 1970s and 1980s, as Alaskan fish numbers dipped, fishermen and government leaders sought more active stock enhancement techniques. In 1976 and 1983–84, the Alaska Department of Fish and Game sent officials to Hokkaido to explore Japanese practices of chum cultivation and hatchery organization (Kron 1985; Moberly and Lium 1977). Illustrating that development does not always flow from the "West to the rest," Alaskans embarked on large-scale hatchery cultivation partially inspired by Japanese models (McNeil 1980, 18).

12 Land-use practices such as clear-cut logging (which produces sediments that smother gravel beds and warm stream temperatures) and mainstem dams (which impede fish passage) are well-known problems for stream-spawning salmon in the Columbia River. However, for hatchery fish, most of which are produced in lower river facilities, the loss of estuary feeding areas is a major issue for which hatcheries do not compensate.

13 In this recounting of differential successes, variations in ocean conditions across the Pacific should also be considered. Differences in ocean conditions may also have contributed to the dramatically divergent return rates of hatchery fish in these regions.

14 In 1991, a National Marine Fisheries document declared that hatchery salmon should not count as salmon under the Endangered Species Act.

According to the policy, "The key is the link between a 'species' and its native habitat, and this link is broken when fish are moved from one ecosystem to another" (Waples 1991, 18–19). For the document's authors, hatchery salmon, whose link to a specific spawning stream was no longer intact, did not represent "an important component in the evolutionary legacy of the species" (12). This statement had major legal and management implications.

15 In the US Pacific Northwest, most hatcheries mark their fish by removing a small fatty fin, called the adipose fin, thus making it visually apparent if a fish is of hatchery origin. Every year, about fifty million juvenile salmon on the US Pacific Coast are also given internal coded wire tags that contain data about their hatchery rearing history (US Fish and Wildlife Service n.d.).

16 For more on these select area fisheries systems, see Columbia River Fish Working Group (2008). Furthermore, while this section has focused on the protection of wild fish relations, US Northwest salmon policies also include ecologically focused activities such as carcass planting, where the bodies of hatchery salmon are placed in streams to improve their nutrition, something that is not a routine part of Japanese salmon management.

17 Overall, Japan's statutes for the conservation of endangered species are much more limited than those of the United States, with no legal mechanisms for citizens to force action. Fish codes are even more limited in that they focus on sustainable catches, not conservation, and delegate most management to fisheries cooperatives (Takahashi 2009).

18 For comparison, around 25 percent of salmon harvested in Alaska in 2019 were of hatchery origin (Welch 2020).

19 See Nagata et al. (2012) for descriptions of changes in Hokkaido salmon management in this period.

20 See Morita (2019), who also discusses the effects of fishing pressure on the diversity of salmon populations along with other risks in relation to climate change. Tillotson et al. (2019) discuss how hatcheries seem to reduce the ability of salmon to cope with warming temperatures from a Northern American context. See also Kitada and Kishino (2019), with the caveat that this study has not been peer reviewed and should thus be seen primarily as an indication of concern and research interests.

21 See graph in Morita (2014, 7).

22 For a Japanese research group's take on these issues, see Kaeriyama et al. (2012).

23 Although this chapter focuses on the United States, it is worth noting that Canada established a formal Wild Salmon Policy in 2005.

1 See also the mention of Ainu children playing cowboys and Indians in Dubreuil (2007).

2 Indigenous scholars have widely analyzed such dynamics. For one well-known example, see Deloria (1969).

3 See Howell (2004) for effects of assimilation policies.

4 This chapter does not intend to make claims about Ainu identity, as it emerges out of research specifically on relations to salmon rather than long-term collaborations with Ainu communities. Furthermore, the subsequent overview of Ainu-salmon relations draws on lines of archeological and historical research that are themselves contested and entangled with webs of problematic comparisons (Kondo and Swanson 2020). It offers one possible reading of a selection of sources but does not intend to be definitive, as various Ainu people may want to narrate these histories in other ways.

5 Ainu peoples are diverse and have deep ties to multiple places, including those currently termed Sakhalin and the Kurils. While this chapter's overview of Ainu-salmon relations focuses on Hokkaido, where Ainu communities were also very different across the island's regions, the Ainu communities with ties to these other islands have their own specific histories, as well as interactions with settler colonialisms (in some cases Russian, as well as Japanese).

6 A distinct set of culture and practices with continuities into the present—referred to as Ainu culture—emerged around this time, so I use the terms *Ainu* and *Ainu Mosir*—the Ainu name for the island—from here onward.

7 This paragraph is based on Segawa (2007) and personal communication with Segawa.

8 See also Iewallen (2016) on the history of Ainu repression and resistance in eastern Hokkaido.

9 For more on this subcontracting system, see Hokkaido/Tohoku Rekishi Kenkyūkai (1998).

10 The information in this paragraph and the subsequent two is largely from interviews with museum staff and scholars in Hokkaido, but see also Kayano (2004, 16).

11 This detail about salmon and salt comes from Segawa, personal communication, April 2010.

12 Although salmon from Ainu Mosir were predominately consumed by poorer people, partially fermented salmon produced in northern Honshu's Niigata region were a delicacy eaten primarily by the upper classes. Tokohu residents sent their own salmon to the tables of Edo

elites, while they themselves ate the tougher, imported Hokkaido salmon (Segawa personal communication).

13　Matstumae officials may have attempted to limit Ainu agriculture during this period to force Ainu into increased trade dependency (Walker 2001, 85–87).

14　Morris-Suzuki (1994, 1996) and Howell (1994, 2004) have written extensively about Japanese state projects toward Ainu people.

15　The herring industry was equally important at this time (Howell 1995).

16　In his 1912 English language book *The Japanese Nation*, Nitobe wrote, "As they are now found, they have not yet emerged from the Stone Age, possessing no art beyond a primitive form of horticulture, being ignorant even of the rudest pottery. Their fate resembles the fate of your American Indians, though they are much more docile in character" (quoted in Harrison 2009, 98).

17　In addition to the ban on salmon fishing, female lip tattoos and poison-tipped hunting arrows, both critical parts of Ainu-ness, were also prohibited.

18　Such efforts were explicitly comparative. For example, in 1874, Benjamin Smith Lyman, an American advisor to the Hokkaido colonization commission, recommended that they eliminate predators, such as wolves, by "offering bounties, as is done in other countries" (Hirano 2015, 206). This recommendation became policy.

19　According to Hirano (2015, 204), in 1871 there were 66,618 Ainu people living in Hokkaido and in 1901, fewer than eighteen thousand. These numbers differ somewhat from those of Walker (2001, 182), regarding the mid-nineteenth century, but both point toward profound losses.

20　For more on Ainu relations with other Indigenous and minority people in this period, including in Greenland, Alaska, and China, see Dietz (1999) and Harrison (2014).

21　One women I interviewed explained it as feeling *tokidoki Ainu*, "sometimes Ainu." The work of scholars who identify as Ainu, including Mai Ishihara's autoethnography (Ishihara 2020) and Kanako Uzawa's descriptions of Ainu youth (Uzawa 2020; Uzawa and Watson 2020), describe related experiences. See also the extended quotes from Ishihara about her experiences of coming to know herself as Ainu in the postscript of Kosaka (2019, 188–91, 274–75).

22　This resonates with lewallen's (2016) description of a person who identifies as Ainu but also runs a commercial fishery. The man refused to give consent for a ceremonial salmon harvest, due to the economic sensitivity of the issue for commercial fisheries, stating, "We can't allow Ainu traditional fishing in our river" (13).

23 For one history of Ainu first salmon ceremonies, see Iwasaki-Goodman and Nomoto (2001).

24 The 1997 law finally replaced the 1899 Ainu protection law that designated the Ainu as "former natives." Although the 1997 law eliminated the worst discriminatory language and provided funding for projects related to Ainu language, arts, and culture, it did little to address economic or rights issues.

25 For an overview of Ainu fishing rights, see Ichikawa (2001).

26 *Ekashi* is an Ainu honorific for male elders. At Hatakeyama-ekashi's request, it is used here instead of the Japanese honorific *-san*.

27 See Uzawa and Watson (2020) for an ethnographic description of Ainu-*wajin* collaborations and their importance within projects for Ainu resurgence. They describe a university group where students with and without Ainu heritage learn about and enact Ainu practices, such as dances, together.

28 This mixed outcome resembles that of the first court ruling that recognized Ainu rights in 1997, in response to the construction of Nibutani Dam, which expropriated Ainu landowners. While the ruling recognized Ainu rights, by that point, the dam had long since been built. See Maruyama (2012).

29 Two key legal rulings were the Belloni decision in 1969 and the Boldt decision in 1974. For one history of fish-in activism, see Shreve (2009).

30 Translation by author from the Japanese provided in Kosaka (2019).

31 The group was formerly called the Urahoro Ainu Association but changed its name to Rahoro Ainu Nation (Rahoro Ainu Neishon), using the English world *nation* transliterated in katakana, likely pointing to another comparison (Kayaba 2020).

32 They also resonate with other Ainu calls for salmon rights, such as Ukaji (2018).

CODA

1 Pacific salmon each have three pairs of otoliths. The largest, the sagittae (about 5 mm in diameter) are usually used for analysis and are those described here.

2 Phrase borrowed from Anderson (2004).

3 Other animals and plants in Japan are bound up with comparisons. See Tsing (2015) on Japanese forests, Skabelund (2011) on dogs, Miller (2013) on the Ueno Zoo, and B. Walker (2005) on Hokkaido's landscapes.

4 Tyrrell (1991) illustrates how American exceptionalism has shaped scholarly approaches to American history in addition to popular narratives.

5 See Tyrrell (1999) and Stoler (2006) for analyses of how the United States has been made through transnational projects characterized by comparative endeavors. Hathaway (2013) also documents how the US feminist movement was deeply inspired by stories of Chinese revolutions, but these influences are almost never mentioned in any histories of US feminism.

WORKS CITED

Abrams, Richard M. 1989. "The US Military and Higher Education: A Brief History." *Annals of the American Academy of Political and Social Science* 502, no. 1 (March): 15–28.

Abu-Lughod, Lila. 1991. "Writing against Culture." In *Recapturing Anthropology: Working in the Present*, edited by Richard G. Fox, 137–54. Santa Fe, NM: School of American Research Press.

Aguayo, Beatriz Cid. 2008. "El 'otro cluster' del salmon: Una mirada a los movimientos sociales." Supported by International Development Research Council of Canada. Biblioteca Nacional de Chile: vag/0/20090217.

Ahrens, H. 1877a. "Oregonshū niokeru keigyo chōzōhō ni kansuru memo/ Aarens shōkai" [A memo on the storage of salmon in Oregon/Ahrens & Co.]. H. Ahrens 007. Kaitakushi gaikokujin kankei shokan mokuroku [Catalog of Hokkaido Development Commission correspondence with foreigners], Hokkaido University Library Northern Studies Collection.

———. 1877b. "Oushū he itaku yūshutsu no kaitakushi sake kanzume ni kansuru hōkokusho/Aarens shōkai" [Report on the consigned export of Hokkaido Development Commission canned salmon to Europe/Ahrens & Co.]. H. Ahrens 008. Kaitakushi gaikokujin kankei shokan mokuroku [Catalog of Hokkaido Development Commission correspondence with foreigners], Hokkaido University Library Northern Studies Collection.

Anderson, Benedict. 1983. *Imagined Communities: Reflections on the Origin and Spread of Nationalism*. London: Verso.

———. 1998. *The Spectre of Comparisons: Nationalism, Southeast Asia, and the World*. New York: Verso.

———. 2005. *Under Three Flags: Anarchism and the Anti-colonial Imagination*. New York: Verso.

Anderson, Virginia DeJohn. 2004. *Creatures of Empire: How Domestic Animals Transformed Early America*. New York: Oxford University Press.

Aoyama, Mami. 2012. "Indigenous Ainu Occupational Identities and the Natural Environment in Hokkaido." In *Politics of Occupation-Centered Practice: Reflections on Occupational Engagement across Cultures*, edited by Nick Pollard and Dikaios Sakellariou, 106–27. Oxford: Wiley.

Appadurai, Arjun. 1990. "Disjuncture and Difference in the Global Cultural Economy." *Theory, Culture & Society* 7, no. 2–3 (June): 295–310.

Araki, Hitoshi, Barry A. Berejikian, Michael J. Ford, and Michael S. Blouin. 2008. "Fitness of Hatchery-Reared Salmonids in the Wild." *Evolutionary Applications* 1 (2): 342–55.

Arase, David, ed. 2005. *Japan's Foreign Aid: Old Continuities and New Directions*. New York: Routledge.

Arnold, David F. 2008. *The Fishermen's Frontier: People and Salmon in Southeast Alaska*. Seattle: University of Washington Press.

Arroyo, Cristián. 2017. "Elaboran primer ranking de empresas con mayor uso de antibióticos en la salmonicultura Chilena." Oceana. https://chile.oceana .org/comunicados/elaboran-primer-ranking-de-empresas-con-mayor-uso -de-antibioticos-en-la.

Asquith, Pamela. 1996. "Japanese Science and Western Hegemonies: Primatology and the Limits Set to Questions." In *Naked Science: Anthropological Inquiry into Boundaries, Power, and Knowledge*, edited by Laura Nader, 239–58. New York: Routledge.

———. 2000. "Negotiating Science: Internationalization and Japanese Primatology." In *Primate Encounters: Models of Science, Gender, and Society*, edited by Shirley C. Strum and Linda Marie Fedigan, 165–83. Chicago: University of Chicago Press.

Augerot, Xanthippe, Dana Nadel Foley, C. Steinback, A. Fuller, N. Fobes, and K. Spencer. 2005. *Atlas of Pacific Salmon: The First Map-Based Status Assessment of Salmon in the North Pacific*. Berkeley: University of California Press.

Barrionuevo, Alexei. 2009. "Chile's Antibiotics Use on Salmon Farms Dwarfs That of a Top Rival's." *New York Times*, July 27, sec. International/ Americas.

Basso, Keith. 1996. *Wisdom Sits in Places: Landscape and Language among the Western Apache*. Albuquerque: University of New Mexico Press.

Batchelor, John. 1902. *Sea-Girt Yezo: Glimpses at Missionary Work in North Japan*. London: Church Missionary Society.

Bauduin, Albertus Johannes. 1879. "Nemurosan sake kanzume no hanro mikomi nitsuki kaitō/Bōdouin" [Correspondence on prospects for sales routes of canned salmon from Nemuro/Bauduin]. Bauduin 045. Kaitakushi gaikokujin kankei shokan mokuroku [Catalog of Hokkaido Development

Commission correspondence with foreigners], Hokkaido University Library Northern Studies Collection.

Beacham, Terry D., Shunpei Sato, Shigehiko Urawa, Khai D. Le, and Michael Wetklo. 2008. "Population Structure and Stock Identification of Chum Salmon (*Oncorhynchus keta*) from Japan Determined by Microsatellite DNA Variation." *Fisheries Science* 74 (5): 983–94.

Befu, Harumi. 1984. "Civilization and Culture: Japan in Search of Identity." *Senri Ethnological Studies* 16 (December): 59–75.

———. 1996. "Watsuji Tetsurō's Ecological Approach." In *Japanese Images of Nature*, edited by Pamela Asquith and Arne Kalland, 106–20. London: Curzon.

Bestor, Theodore C. 2004. *Tsukiji*. Berkeley: University of California Press.

Borie, Adrian Dufflocq. 1981. "Introducción del salmon Pacífico en Chile: Primera etapa programa de investigaciones." Ministerio de Economía, Fomento y Reconstrucción. Chile: Fundación Chile.

Bottom, Daniel L. 1997. "To Till the Water—A History of Ideas in Fisheries Conservation." In *Pacific Salmon & Their Ecosystems*, 569–97. Boston: Springer.

Brophy, Laura S., Correigh M. Greene, Van C. Hare, Brett Holycross, Andy Lanier, Walter N. Heady, Kevin O'Connor, Hiroo Imaki, Tanya Haddad, and Randy Dana. 2019. "Insights into Estuary Habitat Loss in the Western United States Using a New Method for Mapping Maximum Extent of Tidal Wetlands." *PloS One* 14, no. 8 (August): e0218558.

Caple, Zachary. 2017. "Holocene in Fragments: A Critical Landscape Ecology of Phosphorus in Florida." PhD diss., University of California, Santa Cruz.

Capron, Horace M. 1884. *Memoirs of Horace Capron.* Vol. 1, *Autobiography.* Special Collections, National Agricultural Library, Beltsville, MD.

Carlson, Lew. 1989. "Giant Patagonians and Hairy Ainu: Anthropology Days at the 1904 St. Louis Olympics." *Journal of American Culture* 12 (3): 19–26.

Carroll, J. D. 1877. "Sake kunsei shishoku hōkoku, Kyarorusha" [Report on taste-testing smoked salmon, Carroll & Co.]. 0C00987000000000. Kaitakushi gaikokujin kankei shokan mokuroku [Catalog of Hokkaido Development Commission correspondence with foreigners], Hokkaido University Library Northern Studies Collection.

Cassidy, Rebecca, and Molly Mullin. 2007. *Where the Wild Things Are Now: Domestication Reconsidered.* Oxford: Berg.

CEMiPoS. 2020. "Ainu Gather for Kamuycepnomi in Monbetsu (Part 1)." https://cemipos.org/kamuycepnomi-2020/.

CFAJ (Canned Foods Association of Japan). 1934. *Marine Foods Canning Industry in Japan.* Tokyo: Canned Foods Association of Japan.

Cha, Elizabeth. 2004. "The 15 Colors of Salmon." *Wired*. www.wired.com /2004/02/the-15-colors-of-salmon/.

Chacko, Xan Sarah. 2018. "When Life Gives You Lemons: Frank Meyer, Authority, and Credit in Early Twentieth-Century Plant Hunting." *History of Science* 56, no. 4 (July): 432–69.

Choi, Charles. 2008. "Tierra del Fuego: The Beavers Must Die." *Nature News* 453, no. 7198 (June 18): 968.

Choy, Timothy K. 2011. *Ecologies of Comparison: An Ethnography of Endangerment in Hong Kong*. Durham, NC: Duke University Press.

City Population. 2021. "Puerto Montt." www.citypopulation.de/en/chile/mun /admin/llanquihue/10101__puerto_montt/.

Clark, William Smith. 1877a. "Sake kunsei narabi kanzume no hinpyō, hokkaidō no sake masu zōshoku, kanzume yūshutsu no shinkōsaku, kibeigo no chōsa no koto/Kurāku" [Evaluation of smoked and canned salmon, on hatchery policies for Hokkaido salmon and trout and the promotion of canned exports, an investigation after returning to the United States]. Clark, William Smith 065. Kaitakushi gaikokujin kankei shokan mokuroku [Catalog of Hokkaido Development Commission correspondence with foreigners], Hokkaido University Library Northern Studies Collection.

———. 1877b. "Koronbiaka sake gyogyō narabi ni kanzume seizōgyō ni kansuru yobi hôkoku/Kurāku (San furanshisuko)" [Preparatory report on Columbia River salmon fishing and canning/Clark (San Francisco)]. Clark, William Smith 088. Kaitakushi gaikokujin kankei shokan mokuroku [Catalog of Hokkaido Development Commission correspondence with foreigners], Hokkaido University Library Northern Studies Collection.

Clifford, James. 1997. *Routes: Travel and Translation in the Late Twentieth Century*. Cambridge, MA: Harvard University Press.

Clifford, James, and George E. Marcus, eds. 1986. *Writing Culture: The Poetics and Politics of Ethnography*. Berkeley: University of California Press, 1986.

Columbia River Fish Working Group. 2008. "Selective Fisheries." www.co .clatsop.or.us/sites/default/files/fileattachments/fisheries/page/521/selective _fishingoct08.pdf.

Cone, Joseph, and Sandy Ridlington. 2000. *The Northwest Salmon Crisis: A Documentary History*. Corvallis: Oregon State University Press.

Cortazzi, Hugh. 2000. *Collected Writings of Sir Hugh Cortazzi*. Tokyo: Japan Library.

Crawford, Gary W., and Masakazu Yoshizaki. 1987. "Ainu Ancestors and Prehistoric Asian Agriculture." *Journal of Archaeological Science* 14, no. 2 (March): 201–13.

Criddle, Keith, and Ikutaro Shimizu. 2014. "The Economic Importance of Wild Pacific Salmon." In *Salmon*, edited by Patrick T. K. Woo and Donald J. Noakes, 269–306. New York: Nova Science Publishers.

Cronon, William. 1991. *Nature's Metropolis: Chicago and the Great West*. New York: Norton.

Crutzen, P. J., and E. F. Stoermer. 2000. "Anthropocene." *Global Change Newsletter* 41:17–18.

Cunningham, Curry J., Peter A. H. Westley, and Milo D. Adkison. 2018. "Signals of Large Scale Climate Drivers, Hatchery Enhancement, and Marine Factors in Yukon River Chinook Salmon Survival Revealed with a Bayesian Life History Model." *Global Change Biology* 24, no. 9 (May): 4399–416.

Cwiertka, Katarzyna. 2002. "Popularizing a Military Diet in Wartime and Postwar Japan." *Asian Anthropology* 1 (1): 1–30.

———. 2006. *Modern Japanese Cuisine: Food, Power and National Identity*. London: Reaktion.

Czerwien, Christy Anne. 2011. "'Boys Be Ambitious!' The Moral Philosophy of William Smith Clark and the Creation of the Sapporo Band." Master's thesis, University of Pittsburgh.

Dauvergne, Peter. 1997. *Shadows in the Forest: Japan and the Politics of Timber in Southeast Asia*. Cambridge, MA: MIT Press.

Deloria, Vine. 1969. *Custer Died for Your Sins: An Indian Manifesto*. Norman: University of Oklahoma Press.

DeNies, Ramona. 2019. "In Hokkaido, the Dams Finally Come Down." Wild Salmon Center, November 12. https://wildsalmoncenter.org/2019/11/12/in -hokkaido-the-dams-finally-come-down/.

Dietz, Kelly. 1999. "Ainu in the International Arena." In *Ainu: Spirit of a Northern People*, edited by William W. Fitzhugh and Chisato O. Dubreuil, 359–65. Washington, DC: Arctic Studies Center, National Museum of Natural History, Smithsonian Institution, in association with University of Washington Press.

Dower, John W. 1986. *War without Mercy: Race and Power in the Pacific War*. New York: Pantheon.

DSM Animal Nutrition and Health. n.d. "Salmon—DSM Color Fans." www .dsm.com/anh/en_US/solutions/dsm-color-fans/salmon.html.

Dubreuil, Chisato. 2007. "The Ainu and Their Culture: A Critical Twenty-First Century Assessment." *Asia-Pacific Journal: Japan Focus* 5, no. 11 (November): Article ID 2589. https://apjjf.org/-Chisato-Kitty-Dubreuil/2589/article .html.

Dudden, Alexis. 2005. *Japan's Colonization of Korea: Discourse and Power*. Honolulu: University of Hawai'i Press.

———. 2019. "Nitobe Inazo and the Diffusion of a Knowledgeable Empire." In *Empire and the Social Sciences: Global Histories of Knowledge,* ed. Jeremy Adelman, 111–22. London: Bloomsbury Academic.

Duke, Benjamin C. 2009. *The History of Modern Japanese Education: Constructing the National School System, 1872–1890.* New Brunswick, NJ: Rutgers University Press.

Dunlap, Thomas R. 1997. "Remaking the Land: The Acclimatization Movement and Anglo Ideas of Nature." *Journal of World History* 8 (2): 303–19.

Edmonds, Richard L. 1985. *Northern Frontiers of Qing China and Tokugawa Japan: A Comparative Study of Frontier Policy.* Chicago: University of Chicago Press.

Endoh, Toake. 2009. *Exporting Japan: Politics of Emigration to Latin America.* Champaign: University of Illinois Press.

Esposito, Anthony. 2011. "Chile's Salmon Industry on Pace for Record Sales." *MarketWatch.* www.marketwatch.com/story/chiles-salmon-industry-on -pace-for-record-sales-2011-10-04.

Evans, Owen. 2019. "Authorities Laud Reduction in Antibiotics across Chilean Salmon Farms, However Use Is Still Rife." *Salmon Business.* https:// salmonbusiness.com/authorites-laud-reduction-in-antibiotics-in-chilean -salmon-farming-however-use-is-still-rife/.

Fabian, Johannes. (1983) 2014. *Time and the Other: How Anthropology Makes Its Object.* New York: Columbia University Press.

Ferguson, Will. 1998. *Hokkaido Highway Blues: Hitchhiking Japan.* New York: Soho Press.

Finn, Dallas. 1995. *Meiji Revisited: The Sites of Victorian Japan.* New York: Weatherhill.

Foster, David. 2002. "In Oregon, Hatcheries Spawn a Salmon Struggle." *Washington Post.* www.washingtonpost.com/archive/politics/2002/04/21 /in-oregon-hatcheries-spawn-a-salmon-struggle/41a27d1a-f02b-4921 -a045-63784a4dc0f3/.

Fox, Richard, and Andre Gingrich, eds. 2002. *Anthropology, by Comparison.* New York: Routledge.

FRED (Federal Reserve Economic Data). 2021. "Japan/U.S. Foreign Exchange Rate." http://research.stlouisfed.org/fred2/data/EXJPUS.txt.

Freidberg, Susanne Elizabeth. 2004. *French Beans and Food Scares: Culture and Commerce in an Anxious Age.* Oxford: Oxford University Press.

Freres, Peyre, 1878. "Kaitakushisei sake kanzume hinpyō (eiyaku)/Peerusha" [An evaluation of Hokkaido Development Commission-made canned salmon (English translation)/Peyre Brothers & Co]. Peyre Freres 006. Kaitakushi gaikokujin kankei shokan mokuroku [Catalog of Hokkaido

Development Commission correspondence with foreigners], Hokkaido University Library Northern Studies Collection.

Frey, Christopher J. 2007. "Ainu Schools and Education Policy in Nineteenth-Century Hokkaido, Japan." PhD diss., Indiana University.

Fujita, Fumiko. 1994. *American Pioneers and the Japanese Frontier: American Experts in Nineteenth-Century Japan.* Westport, CT: Greenwood Press.

Gad, Christopher, and Casper Bruun Jensen. 2016. "Lateral Comparisons." In *Practicing Comparison: Logics, Relations, Collaborations,* edited by Joe Deville, Michael Guggenheim, and Zuzana Hrdličková, 189–219. Manchester: Mattering Press.

Gell, Alfred. 1996. "Vogel's Net: Traps as Artworks and Artworks as Traps." *Journal of Material Culture* 1, no. 1 (March): 15–38.

Gellner, Ernest. 1983. *Nations and Nationalism.* Ithaca, NY: Cornell University Press.

GHQ/SCAP Records. 1950. "Information Program—Fisheries Cooperatives." RG 331 National Archives and Records Service, Box 5337, Folder 8. National Diet Library, Tokyo, Japan.

Gluck, Carol. 1997. "Meiji for Our Time." In *New Directions in the Study of Meiji Japan,* edited by Helen Hardacre and Adam Kern, 11–28. Leiden: Brill.

———. 2011. "The End of Elsewhere: Writing Modernity Now." *American Historical Review* 116, no. 3 (June): 676–87.

Grunow, Tristan R., Fuyubi Nakamura, Katsuya Hirano, Mai Ishihara, ann-elise lewallen, Sheryl Lightfoot, Mayunkiki, Danika Medak-Saltzman, Terri-Lynn Williams-Davidson, and Tomoe Yahata. 2019. "Hokkaidō 150: Settler Colonialism and Indigeneity in Modern Japan and Beyond." *Critical Asian Studies* 51, no. 4 (October): 597–636.

Hara, Yoshio. 1977. "From Westernization to Japanization: The Replacement of Foreign Teachers by Japanese Who Studied Abroad." *Developing Economies* 15, no. 4 (December): 440–61.

Haraway, Donna. 2003. *The Companion Species Manifesto: Dogs, People, and Significant Otherness.* Chicago: Prickly Paradigm Press.

———. 2008. *When Species Meet.* Minneapolis: University of Minnesota Press.

———. 2015. "Anthropocene, Capitalocene, Plantationocene, Chthulucene: Making Kin." *Environmental Humanities* 6, no. 1 (May): 159–65.

Harrison, John. 1951. "The Capron Mission and the Colonization of Hokkaido, 1868–1875." *Agricultural History* 25 (3): 135–42.

Harrison, Scott. 2009. "The Indigenous Ainu of Japan at the Time of the Åland Settlement." In *Northern Territories, Asia-Pacific Regional Conflicts and the Åland Experience,* 95–105. London: Routledge.

———. 2014. "The Cold War, the San Francisco System, and Indigenous Peoples." In *The San Francisco System and Its Legacies*, edited by Kimie Hara, 203–19. New York: Routledge.

Hasegawa, Koh. 2020. "Invasions of Rainbow Trout and Brown Trout in Japan: A Comparison of Invasiveness and Impact on Native Species." *Ecology of Freshwater Fish* 29, no. 3 (January): 419–28.

Hathaway, Michael. 2013. *Environmental Winds: Making the Global in Southwest China*. Berkeley: University of California Press.

Hébert, Karen. 2010. "In Pursuit of Singular Salmon: Paradoxes of Sustainability and the Quality Commodity." *Science as Culture* 19 (4): 553–81.

———. 2015. "Enduring Capitalism: Instability, Precariousness, and Cycles of Change in an Alaskan Salmon Fishery." *American Anthropologist* 117, no. 1 (February): 32–46.

Hennessey, John L. 2020. "A Colonial Trans-Pacific Partnership: William Smith Clark, David Pearce Penhallow and Japanese Settler Colonialism in Hokkaido." *Settler Colonial Studies* 10 (1): 54–73.

Hilborn, Ray. 1992. "Hatcheries and the Future of Salmon in the Northwest." *Fisheries* 17, no. 1 (January): 5–8.

Hirano, Katsuya. 2015. "Thanatopolitics in the Making of Japan's Hokkaido: Settler Colonialism and Primitive Accumulation." *Critical Historical Studies* 2, no. 2 (Fall): 191–218.

Hogan, Jackie. 2003. "The Social Significance of English Usage in Japan." *Japanese Studies* 23, no. 1 (August): 43–58.

Hokkaido National Fisheries Research Institute, Japan Fisheries Research and Education Agency, National Research and Development Agency. 2020. "Changes in Salmon Release and Return Numbers Including Return Percentages." http://salmon.fra.affrc.go.jp/zousyoku/ok_relret.html.

Hokkaido Prefectural Government. 1968. *Foreign Pioneers: A Short History of the Contribution of Foreigners to the Development of Hokkaido*. Sapporo: Hokkaido Prefectural Government.

Hokkaido/Tohoku Rekishi Kenkyūkai [Hokkaido/Tohoku historical studies association]. 1998. *Basho ukeoisei to ainu: Kinsei ezochishi no kōchiku wo mezashite: Sapporo shinpojiumu* [Ainu and the location contract system: Toward the construction of early modern Ezo history: Sapporo Symposium]. Hokkaidô Shuppan Kikaku Sentaa [Hokkaido publishing planning center].

Hosaka, Kazuyoshi. 1993. "Similar Introduction and Incorporation of Potato Chloroplast DNA in Japan and Europe." *Japanese Journal of Genetics* 68, no. 1 (January): 55–61.

Hosono, Akio. 2010. *Nanbei chiri wo sake yûshutsu daikoku ni kaeta nihonjin-tachi: Zero kara sangyō wo zōshutsu kokusai kyōryoku no kiroku* [The Japanese who made Chile into a salmon exporting power: A record of international cooperation in developing hatcheries from zero]. Daiya-mondo biggusha, Daiyamondosha [Daimon big company, Daimond company].

Howell, David. 1994. "Ainu Ethnicity and the Boundaries of the Early Modern Japanese State." *Past & Present*, no. 142 (February 1), 69–93.

———. 1995. *Capitalism from Within: Economy, Society, and the State in a Japanese Fishery*. Berkeley: University of California Press.

———. 1999. "The Ainu and the Early Modern Japanese State." In *Ainu: Spirit of a Northern People*, edited by William W. Fitzhugh and Chisato O. Dubreuil, 96–101. Washington, DC: Arctic Studies Center, National Museum of Natural History, Smithsonian Institution, in association with University of Washington Press.

———. 2004. "Making 'Useful Citizens' of Ainu Subjects in Early Twentieth-Century Japan." *Journal of Asian Studies* 63 (1): 5–29.

Hutchinson, William E. 1951. *History of the Nonmilitary Activities of the Occupation of Japan 1945–1950*. Vol. 14, *Natural Resources—Part B, Fisheries*, by the Supreme Commander for the Allied Powers, Civil Historical Section. GHQ/SCAP Records, Office of Civil Property Custodian. SCAP Monograph Drafts, 1945–51. Box 3678, Folder 10, Monograph Draft #42. Materials on the Allied Occupation of Japan, Japanese National Diet Library, Tokyo.

Ichikawa, Morihiro. 2001. "Understanding the Fishing Rights of the Ainu of Japan: Lessons Learned from American Indian Law, the Japanese Constitution, and International Law." *Colorado Journal of International Environmental Law and Policy* 12 (2): 245–301.

Ichiryūkai [The Ichiryū Association]. 1987. *Itō Kazutaka to tsunagaru hitobito* [The people connected to Itō Kazutaka]. Tokyo: Itô Nakashi.

Indigenous Peoples Rights International. 2020. "Criminalising Rituals and Traditional Occupations: The Struggle of Ainu in Japan, a Century Hence." https://indigenousrightsinternational.org/news-and-events/news -and-features/criminalising-rituals-and-traditional-occupations-the -struggle-of-ainu-in-japan-a-century-hence.

Inoue, Takutoshi. 2008. "Japanese Students in England and the Meiji Government's Foreign Employees (Oyatoi): The People Who Supported Modernisation in the Bakumatsu-Early Meiji Period." Discussion Paper Series, No. 40. School of Economics, Kwansei Gakuin University. https://core.ac .uk/reader/143634920.

International Museum of the Horse. n.d. "Japanese Native Horses." http://imh
.org/exhibits/online/breeds-of-the-world/asia/japanese-native-horses/.

IntraFish Media. 2010. "Salmon Is Japan's Favorite Fish." *IntraFish Online*,
May 26. https://www.intrafish.com/news/salmon-is-japan-rsquo-s-favorite
-fish/1-1-624886.

Irish, Ann B. 2009. *Hokkaido: A History of Ethnic Transition and Development
on Japan's Northern Island*. Jefferson, NC: McFarland.

Ishihara, Mai. 2020. *Chinmoku no jitekiminzokushi—Sairento-Ainu no itami
to kyūsai no monogatari* [Autoethnography of silence: The story of the pain
of silent Ainu and their care]. Sapporo: Hokkaido University Press.

Ishikawa, Noboru, and Mayumi Ishikawa. 2013. "Global Timber Connections:
A Critical Look at Forests in Japan and Southeast Asia." Paper presented at
American Anthropological Association (AAA) 2013 Annual Meeting
(November 20–24), Chicago.

Ishikawa, Takuboku. 1967. *Hajimete mitaru Otaru* [First sight of Otaru].
Nihon Bungaku Zenshu. Vol. 12. Tokyo: Shueisha. Digital version at www
.aozora.gr.jp/cards/000153/files/812_20595.html.

Ito, Kazutaka. 1890. "Hokkaidōchō dainibu suisanka. Beikoku gyogyō chōsa
fukumeisho. Hokkaidōchō dainibu suisanka" [Hokkaido agency second
fisheries section. Reply to the American fishing industry investigations].
National Diet Library Digital Collections. https://dl.ndl.go.jp/info:ndljp/pid
/842769.

Ivings, Steven, and Datong Qiu. 2019. "China and Japan's Northern Frontier:
Chinese Merchants in Nineteenth-Century Hokkaido." *Canadian Journal
of History* 54 (3): 286–314.

Ivy, Marilyn. 1995. *Discourses of the Vanishing: Modernity, Phantasm, Japan*.
Chicago: University of Chicago Press.

Iwama, Kazuto. 2009. "Cultivation of Field Crops." In *Agriculture in Hokkaido*,
edited by Kazuto Iwama, Masashi Ohara, Hajime Araki, Toshihiko
Yamada, Hiroki Nakatsuji, Takashi Kataoka, and Yasutaka Yamamoto.
Sapporo: Hokkaido University Press.

Iwasaki-Goodman, Masami, and Masahiro Nomoto. 2001. "Revitalizing the
Relationship between Ainu and Salmon: Salmon Rituals in the Present."
Senri Ethnological Studies 59:27–46.

IWGIA (International Workgroup for Indigenous Affairs). 2021. "The Indigenous World 2021: Japan." www.iwgia.org/en/japan/4226-iw-2021-japan
.html#_ftn20.

Jaksic, Fabian M., J. Agustín Iriarte, Jaime E. Jiménez, and David R. Martínez.
2002. "Invaders without Frontiers: Cross-Border Invasions of Exotic
Mammals." *Biological Invasions* 4, no. 1–2 (March 1): 157–73.

Japan Bird Research Association. 2010. "Blakiston's Fish Owl." *Bird Research News* 7 (2): 4–5.

Japan Ministry of Agriculture, Forestry and Fisheries, Hokkaido Agricultural Policy Office. 2020. Gurafu de Miru Hokkaido Gyogyō. Hokkaido Fisheries Presented in Graphs (September). www.maff.go.jp/hokkaido/toukei/kikaku /gurafu_gaiyou/gyogyou2010/attach/pdf/gyogyo2013–55.pdf.

Jensen, Casper B., and Atsuro Morita. 2017. "Introduction: Minor Traditions, Shizen Equivocations, and Sophisticated Conjunctions." *Social Analysis* 61, no. 2 (June): 1–14.

Jensen, Casper B., Barbara Herrnstein Smith, G. E. R. Lloyd, Martin Holbraad, Andreas Roepstorff, Isabelle Stengers, Helen Verran, et al. 2011. "Introduction: Contexts for a Comparative Relativism." *Common Knowledge* 17, no. 1 (January): 1–12.

Kaeriyama, Masahide. 1989. "Aspects of Salmon Ranching in Japan." *Physiological Ecology Japan*, special volume 1: 625–38.

Kaeriyama, Masahide, Hyunju Seo, Hideaki Kudo, and Mitsuhiro Nagata. 2012. "Perspectives on Wild and Hatchery Salmon Interactions at Sea, Potential Climate Effects on Japanese Chum Salmon, and the Need for Sustainable Salmon Fishery Management Reform in Japan." *Environmental Biology of Fishes* 94, no. 1 (May): 165–77.

Kamuycep Project Research Group. 2021. *Kamuycep dokuhon: Hokkaidō no atarashii sake kanri* [Kamuycep textbook: Hokkaido's new salmon management]. Sapporo: Sapporo Freedom School "Yu."

Kataoka, Takashi. 2009. "Agricultural Machinery Technology." In *Agriculture in Hokkaido*, edited by Kazuto Iwama, Masashi Ohara, Hajime Araki, Toshihiko Yamada, Hiroki Nakatsuji, Takashi Kataoka, and Yasutaka Yamamoto, 6-1–6-9. Sapporo: Hokkaido University Press.

Katz, Jorge. 2006. "Salmon Farming in Chile." In *Technology, Adaptation, and Exports: How Some Developing Countries Got It Right*, edited by Vandana Chandra, 193–223. Washington, DC: World Bank Publications.

Kawagoe, Toshihiko. 1999. "Agricultural Land Reform in Postwar Japan: Experiences and Issues." *Policy Research Working Papers*, vol. 2111 (November). Washington, DC: World Bank Publications.

Kawakami, Kiyoshi. 1906. "A Japanese on Japan; Second Review of Alfred Stead's Book from the Point of View of One of Its Subjects—Japan Embarrassed by Overpraise." *New York Times Saturday Review of Books*, July 21.

Kayaba, Yūta. 2020. "Ainu senjyū-ken soshō 'Ōkina tenkan-ten ni' 'Giron fukamareba.'" [Ainu indigenous rights lawsuit: "A big turning point" "If the discussion deepens"]. *Asahi Shinbun*. August 18. www.asahi.com /articles/ASN8K7D8GN8KIIPE00Q.html.

Kayano, Shigeru. 1994. *Our Land Was a Forest: An Ainu Memoir.* Boulder, CO: Westview Press.

———. 2004. *The Ainu: A Story of Japan's Original People.* Translated by Peter Howlett and Richard McNamara. Boston: Tuttle.

Kerr, Alex. 2001. *Dogs and Demons: Tales from the Dark Side of Japan.* New York: Macmillan.

Kikuchi, Toshihiko. 1999. "Ainu Ties with Ancient Cultures of Northeast Asia." In *Ainu: Spirit of a Northern People,* edited by William W. Fitzhugh and Chisato O. Dubreuil, 47–51. Washington, DC: Arctic Studies Center, National Museum of Natural History, Smithsonian Institution, in association with University of Washington Press.

Kirksey, Eben, and Stefan Helmreich. 2010. "The Emergence of Multispecies Ethnography." *Cultural Anthropology* 25 (4): 545–76.

Kitada, Shuichi. 2014. "Japanese Chum Salmon Stock Enhancement: Current Perspective and Future Challenges." *Fisheries Science* 80, no. 2 (January): 237–49.

Kitada, Shuichi, and Hirohisa Kishino. 2019. "Fitness Decline in Hatchery-Enhanced Salmon Populations Is Manifested by Global Warming." bioRxiv pre-print server, 828780. doi: https://doi.org/10.1101/828780.

Kitayama, Shinobu, Keiko Ishii, Toshie Imada, Kosuke Takemura, and Jenny Ramaswamy. 2006. "Voluntary Settlement and the Spirit of Independence: Evidence from Japan's Northern Frontier." *Journal of Personality and Social Psychology* 91 (3): 369–84.

Klubock, Thomas. 2014. *La Frontera: Forests and Ecological Conflict in Chile's Frontier Territory.* Durham, NC: Duke University Press.

Knapp, Gunnar. 2007. "Implications of Aquaculture for Wild Fisheries: The Case of Alaska Wild Salmon." In *Global Trade Conference on Aquaculture, 29–31 May 2007, Qingdao, China,* edited by Richard Arthur and Jochen Nierentz, 239–45. FAO Fisheries Proceedings 9.

Kobayashi, Tetsuo. 1980. "Salmon Propagation in Japan." In *Salmon Ranching,* edited by J. E. Thorpe, 91–107. New York: Academic Press.

———. 2009. *Nihon sake-masu zōshokushi* [A history of salmon and trout hatcheries in Japan]. Sapporo: Hokkaidō daigaku shuppankai [Hokkaido university publishing society].

Kohara, Toshihiro. 1999. "Foods of Choice." In *Ainu: Spirit of a Northern People,* edited by William W. Fitzhugh and Chisato O. Dubreuil, 202–7. Washington, DC: Arctic Studies Center, National Museum of Natural History, Smithsonian Institution, in association with University of Washington Press.

Kolbert, Elizabeth. 2014. *The Sixth Extinction: An Unnatural History*. London: A&C Black.

Kondo, Norihisa. 1993. "Mammal Fauna and Its Distribution in Hokkaido." In *Biodiversity and Ecology in the Northernmost Japan*, edited by Seigo Higashi, Akira Osawa, and Kana Kanagawa, 76–87. Sapporo: Hokkaido University Press.

Kondo, Shiaki, and Heather Anne Swanson. 2020. "Sake Masu Ron (Salmon Trout Theory) and the Politics of Non-Western Academic Terms." *Sociological Review* 68, no. 2 (April): 435–51.

Kosaka, Yosuke. 2018. "Revival of Salmon Resources and Restoration of a Traditional Ritual of the Ainu, the Indigenous People of Japan." In *Indigenous Efflorescence: Beyond Revitalisation in Sapmi and Ainu Mosir*, edited by Gerald Roche, Hiroshi Maruyama, and Asa Virdi Kroik, 69–78. Acton: Australian National University Press.

———. 2019. *The Ainu and the Japanese: Different Ground Gives Life to Different Spirits*. Kushiro: Fujita Printing Company Excellent Books.

Krämer, Hans Martin. 2008. "'Not Befitting Our Divine Country': Eating Meat in Japanese Discourses of Self and Other from the Seventeenth Century to the Present." *Food and Foodways* 16, no. 1 (March): 33–62.

Kron, Thomas. 1985. "Japan's Salmon Culture Program and Coastal Salmon Fisheries, Report 50." Juneau: Alaska Department of Fish and Game Division of Fisheries, Rehabilitation, Enhancement, and Development. September. www.adfg.alaska.gov/FedAidpdfs/FRED.050.pdf.

Kublin, Hyman. 1959. "The Evolution of Japanese Colonialism." *Comparative Studies in Society and History* 2, no. 1 (October 1): 67–84.

Kuroda, Kiyotaka. 1877a. "Nemurosan sake, Etorofusan masu no kunsei narabi kanzume no hinpyō oyobi beikoku no shijō kyōshi irai (hikae)/Kuroda chōkan (Tokyo)" [Evaluation of smoked and canned Nemuro salmon and Etorofu (southern Sakhalin) trout and a request for information on the American market (memo)/Chief Kuroda (Tokyo)]. Clark, William Smith 061. Kaitakushi gaikokujin kankei shokan mokuroku [Catalog of Hokkaido Development Commission correspondence with foreigners], Hokkaido University Library Northern Studies Collection.

———. 1877b. "Sake kanzume gijutsusha ichimei kōyō irai/ Kuroda chōkan" [A request for employment of one salmon canning technician/Chief Kuroda]. Capron, Horace 039. Clark, William Smith 061. Kaitakushi gaikokujin kankei shokan mokuroku [Catalog of Hokkaido Development Commission correspondence with foreigners], Hokkaido University Library Northern Studies Collection.

Kuwayama, Takami. 2004. *Native Anthropology: The Japanese Challenge to Western Academic Hegemony*. Rosanna: Trans Pacific Press.

Larson, Erik, Zachary Johnson, and Monique Murphy. 2008. "Emerging Indigenous Governance: Ainu Rights at the Intersection of Global Norms and Domestic Institutions." *Alternatives* 33 (1): 53–82.

Latta, Alex, and Beatriz E. Cid Aguayo. 2012. "Testing the Limits: Neoliberal Ecologies from Pinochet to Bachelet." *Latin American Perspectives* 39, no. 4 (March 13): 163–80. https://www.jstor.org/stable/23239012.

Law, John. 2008. "On Sociology and STS." *Sociological Review* 56, no. 4 (November): 623–49.

Lawson, Joseph. 2015. "The Chinese State and Agriculture in an Age of Global Empires, 1880–1949." In *Eco-cultural Networks and the British Empire*, edited by James Beattie, Edward Melillo, and Emily O'Gorman, 44–68. London: Bloomsbury.

Learn, Scott. 2013. "Judge Allows Release of Sandy Hatchery Salmon, but at a Reduced Level." *Oregonian*. Last modified January 10, 2019. www.oregonlive.com/environment/2013/03/judge_allows_release_of_sandy.html.

Lever, Christopher. 1992. *They Dined on Eland: The Story of the Acclimatisation Societies*. London: Quiller Press.

lewallen, ann-elise. 2016. "Signifying Ainu Space: Reimagining Shiretoko's Landscapes through Indigenous Ecotourism." *Humanities* 5, no. 3 (July): 59.

Lewis, Simon L., and Mark A. Maslin. 2015. "Defining the Anthropocene." *Nature* 519, no. 7542 (March): 171–80.

———. 2018. *The Human Planet*. New Haven, CT: Yale University Press.

Lichatowich, Jim. 1999. *Salmon without Rivers: A History of the Pacific Salmon Crisis*. Washington, DC: Island Press.

Lien, Marianne E. 2005. "'King of Fish' or 'Feral Peril': Tasmanian Atlantic Salmon and the Politics of Belonging." *Environment and Planning D: Society and Space* 23 (5): 659–71.

Limerick, Patricia Nelson. 1988. *The Legacy of Conquest: The Unbroken Past of the American West*. New York: Norton.

Liu, Lydia He. 1995. *Translingual Practice: Literature, National Culture, and Translated Modernity–China, 1900–1937*. Redwood City, CA: Stanford University Press.

Lorimer, Jamie. 2017. "The Anthropo-scene: A Guide for the Perplexed." *Social Studies of Science* 47, no. 1 (October): 117–42.

Lu, David J. (1885) 1996. "Good-bye Asia." In *Japan: A Documentary History*, 351–53. Totnes: East Gate Books.

Lu, Sidney Xu. 2019. *The Making of Japanese Settler Colonialism*. Cambridge, UK: Cambridge University Press.

Magnusson, Lars. 2009. *Nation, State and the Industrial Revolution: The Visible Hand*. London: Routledge.

Maki, John McGilvrey. (1996) 2002. *A Yankee in Hokkaido: The Life of William Smith Clark*. Lanham, MD: Lexington Books.

Makino, Mitsutaku, and Hiroyuki Matsuda. 2005. "Co-management in Japanese Coastal Fisheries: Institutional Features and Transaction Costs." *Marine Policy* 29, no. 5 (September): 441–50.

Mannen, H., S. Tsuji, R. T. Loftus, and D. G. Bradley. 1998. "Mitochondrial DNA Variation and Evolution of Japanese Black Cattle (Bos Taurus)." *Genetics* 150, no. 3 (November): 1169–75.

Manzenreiter, Wolfram. 2017. "Living under More Than One Sun: The Nikkei Diaspora in the Americas." *Contemporary Japan* 29, no. 2 (July): 193–213.

Marine Stewardship Council. 2010. "TAB D-001 V2 Enhanced Fisheries— Scope of Application of the MSC Principles and Criteria." www.msc.org /documents/consultations/consultations/scheme-document-review/tab -directives/TAB_D_001_Enhanced_Fisheries_v2.1-changes-tracked.pdf /view.

Marsh, George Perkins. 1855. "The Camel." In *The Ninth Annual Report of the Board of Regents of the Smithsonian Institution for 1854*, Section 14, 98–167. Washington, DC: Beverly Tucker, Senate Printer.

Martin, Irene, and Roger Tetlow. 2011. "Flight of the Bumble Bee: The Columbia River Packers Association and a Century in the Pursuit of Fish." Long Beach, WA: Chinook Observer.

Maruyama, Hiroshi. 2012. "Ainu Landowners' Struggle for Justice and the Illegitimacy of the Nibutani Dam Project in Hokkaido Japan." *International Community Law Review* 14, no. 1 (January): 63–80.

Mason, Michele M. 2005. "Manly Narratives: Writing Hokkaido into the Political and Cultural Landscape of Imperial Japan." PhD diss., University of California, Irvine.

———. 2012a. *Dominant Narratives of Colonial Hokkaido and Imperial Japan: Envisioning the Periphery and the Modern Nation-State*. New York: Palgrave Macmillan.

———. 2012b. "Writing Ainu Out/Writing Japanese In: The 'Nature' of Japanese Colonialism." In *Reading Colonial Japan: Text, Context, and Critique*, edited by Michele Mason and Helen Lee, 33–54. Redwood City, CA: Stanford University Press.

Masterson, Daniel M., and Sayaka Funada-Classen. 2004. *The Japanese in Latin America*. Champaign: University of Illinois Press.

Mathews, Andrew S. 2018. "Landscapes and Throughscapes in Italian Forest Worlds: Thinking Dramatically about the Anthropocene." *Cultural Anthropology* 33, no. 3 (September): 386–414.

Matsuda, Yoshiaki. 2002. "History of Fisheries: Science in Japan." In *Oceanographic History: The Pacific and Beyond*, edited by Keith Rodney Benson and Philip F. Rehbock, 405–16. Seattle: University of Washington Press.

McCormack, Gavan. (1996) 2016. *The Emptiness of Japanese Affluence.* London: Routledge.

McNeil, W. J. 1980. "Salmon Ranching in Alaska." In *Salmon Ranching*, edited by John Thorpe, 13–27. New York: Academic Press.

McNeill, J. R., and Peter Engelke. 2016. *The Great Acceleration: An Environmental History of the Anthropocene since 1945.* Cambridge, MA: Harvard University Press.

Medak-Saltzman, Danika. 2008. "Staging Empire: The Display and Erasure of Indigenous Peoples in Japanese and American Nation Building Projects (1860–1904)." PhD diss., University of California, Berkeley.

———. 2010. "Transnational Indigenous Exchange: Rethinking Global Interactions of Indigenous Peoples at the 1904 St. Louis Exposition." *American Quarterly* 62 (3): 591–615.

Mendez, Ricardo. 1982. "A Program for the Introduction of Pacific Salmon in Southern Chile." Fundación Chile. Fundación Chile Library.

Miles, Richard D., and Frank A. Chapman. 2006. "The Benefits of Fish Meal in Aquaculture Diets." *EDIS IFAS Extension*, FA122, no. 12 (May): 1–7.

Miller, Ian Jared. 2013. *The Nature of the Beasts: Empire and Exhibition at the Tokyo Imperial Zoo.* Vol. 27. Berkeley: University of California Press.

Milstein, Michael. 2017. "NOAA Fisheries Completes Review of Columbia River Hatcheries." NOAA Fisheries. Last modified September 27, 2019. www.fisheries.noaa.gov/press-release/noaa-fisheries-completes-review -columbia-river-hatcheries.

Minamoto, Shoku. 1993. "The Beginnings of Modern Geography in Japan: From the Mid-Nineteenth Century to the 1910s." In *Geographical Studies & Japan*, edited by John Sargent and Richard Wiltshire, 25–38. London: Psychology Press.

Mintz, Sidney Wilfred. 1985. *Sweetness and Power: The Place of Sugar in Modern History.* New York: Penguin.

Miyakoshi, Yasuyuki, Mitsuhiro Nagata, Shuichi Kitada, and Masahide Kaeriyama. 2013. "Historical and Current Hatchery Programs and Management of Chum Salmon in Hokkaido, Northern Japan." *Reviews in Fisheries Science* 21, no. 3–4 (November): 469–79.

Moberly, Stanley A., and Robert Lium. 1977. "Japan Salmon Hatchery Review." *Fisheries* 2 (3): 2–7.

Mohácsi, Gergely, and Atsuro Morita. 2013. "Traveling Comparisons: Ethnographic Reflections on Science and Technology." *East Asian Science, Technology and Society: An International Journal* 7, no. 2 (October): 175–83.

Mol, Annemarie. 2003. *The Body Multiple*. Durham, NC: Duke University Press.

Montgomery, David R. 2003. *King of Fish: The Thousand-Year Run of Salmon*. Boulder, CO: Westview Press.

Moore, Jason W. 2017. "The Capitalocene, Part I: On the Nature and Origins of Our Ecological Crisis." *Journal of Peasant Studies* 44, no. 3 (March): 594–630.

Morita, Kentaro. 2014. "Japanese Wild Salmon Research: Toward a Reconciliation between Hatchery and Wild Salmon Management." *North Pacific Anadromous Fish Commission Newsletter* 35:4–14.

———. 2019. "Earlier Migration Timing of Salmonids: An Adaptation to Climate Change or Maladaptation to the Fishery?" *Canadian Journal of Fisheries and Aquatic Sciences* 76 (3): 475–79.

Morita, Kentaro, Toshihiko Saito, Yasuyuki Miyakoshi, Masa-aki Fukuwaka, Toru Nagasawa, and Masahide Kaeriyama. 2006. "A Review of Pacific Salmon Hatchery Programmes on Hokkaido Island, Japan." *ICES Journal of Marine Science* 63 (7): 1353–63.

Morris-Suzuki, Tessa. 1994. "Creating the Frontier: Border, Identity and History in Japan's Far North." *East Asian History* 7:1–24.

———. 1996. "A Descent into the Past: The Frontier in the Construction of Japanese History." In *Multicultural Japan: Paleolithic to Postmodern*, edited by Donald Denoon, Mark Hudson, Gavan McCormack, and Tessa Morris-Suzuki, 81–94. Cambridge, UK: Cambridge University Press.

———. 1998a. "Becoming Japanese: Imperial Expansion and Identity Crises in the Early Twentieth Century." In *Japan's Competing Modernities: Issues in Culture and Democracy, 1900–1930*, edited by Sharon Minichiello, 157–80. Honolulu: University of Hawai'i Press.

———. 1998b. *Re-inventing Japan: Time, Space, Nation*. Japan in the Modern World. Armonk, NY: M. E. Sharpe.

Nagata, Mitsuhiro, Yasuyuki Miyakoshi, Hirokazu Urabe, Makoto Fujiwara, Yoshitaka Sasaki, Kiyoshi Kasugai, Mitsuru Torao, Daisei Ando, and Masahide Kaeriyama. 2012. "An Overview of Salmon Enhancement and the Need to Manage and Monitor Natural Spawning in Hokkaido, Japan." *Environmental Biology of Fishes* 94, no. 1 (May 1): 311–23.

Nahuelhual, Laura, Alejandra Carmona, Antonio Lara, Cristian Echeverría, and Mauro E. González. 2012. "Land-Cover Change to Forest Plantations: Proximate Causes and Implications for the Landscape in South-Central Chile." *Landscape and Urban Planning* 107, no. 1 (July): 12–20.

Nakagawa, Daisuke. 2009. "Yasei sake no hozen, kakkoku ga jûshi" [Wild salmon conservation, every nation gives it serious consideration]. *Hokkaidō Shinbun*, February 24, yūkan zendō chiban [Evening edition, all-Hokkaido late edition]. Tokushū [special].

Nakayama, Shigeru. 1989. "Independence and Choice: Western Impacts on Japanese Higher Education." *Higher Education* 18, no. 1 (January 1): 31–48.

Nash, Colin E. 2011. *The History of Aquaculture.* Ames, IA: Wiley-Blackwell.

Naylor, Simon. 2000. "Spacing the Can: Empire, Modernity, and the Globalisation of Food." *Environment and Planning A* 32 (9): 1625–40.

Needham, Rodney. 1975. "Polythetic Classification: Convergence and Consequences." *Man* 10, no. 3 (September): 349–69.

New York Times. 1879. "The Isles of the Pacific; A Curious Ethnological Chart. Odd Stories of Race Migration Disclosed—A Map at the American Museum of Natural History—the Race which the Japanese Drove Away." August 9. http://select.nytimes.com/gst/abstract.html?res=F00A11FB385A127B93CBA91783D85F4D8784F9.

Nihon Keizai Shinbun. 2011. "Chiri de sake yōshoku mitsubishi shōji genchi ōte wo kaishū" [Mitsubishi Trading purchases major local breeder of salmon in Chile]. *Nihon Keizai Shinbun*, November 9, 2011.

Nitobe, Inazō. 1893. "The Imperial Agricultural College of Sapporo, Japan." http://catalog.hathitrust.org/api/volumes/oclc/13323053.html.

Noguchi, Fumiko. 2017. "A Radical Approach from the Periphery: Informal ESD through Rights Recovery for Indigenous Ainu." In *Educating for Sustainability in Japan*, edited by Jane Singer, Tracey Gannon, Fumiko Noguchi, and Yoko Mochizuki, 201–15. London: Routledge.

Northwest Power and Conservation Council. n.d. "Hatcheries." www.nwcouncil.org/history/hatcheries.asp.

OCED (Organisation for Economic Cooperation and Development). 2021. "Fisheries and Aquaculture in Japan." Review of Fisheries Country Notes (January). https://www.oecd.org/agriculture/topics/fisheries-and-aquaculture/documents/report_cn_fish_jpn.pdf.

Ogden, Laura. 2021. *Loss and Wonder at the World's End.* Durham, NC: Duke University Press.

Ohnuki-Tierney, Emiko. 1993. *Rice as Self: Japanese Identities through Time.* Princeton, NJ: Princeton University Press.

Oka, Takashi. 1981. "Hokkaido: Japan's Last Frontier." *Christian Science Monitor*, October 15. www.csmonitor.com/1981/1015/101550.html.

Okamoto, Yasutaka. 2009. "Kita taiheiyō to nihon ni okeru sakemasurui no shigen to zōshoku" [Salmon and trout resources and hatchery production in Japan and the North Pacific]. *SALMON* jōhō *Bulletin of the National Salmon Resources Center* 3 (January): 24–25.

Ono, Yugo. 1999. "Ainu Homelands: Natural History from Ice Age to Modern Times." In *Ainu: Spirit of a Northern People*, edited by William W. Fitzhugh and Chisato O. Dubreuil, 32–38. Washington, DC: Arctic Studies Center, National Museum of Natural History, Smithsonian Institution, in association with University of Washington Press.

O'Ryan, Raúl, Mario Niklitschek, Edwin Niklitschek, Nicolo Gligo, and Andrés Ulloa. 2010. "Trade Liberalization, Rural Poverty and the Environment: A Case Study of the Forest and Salmon Sectors in Chile." In *Vulnerable Places, Vulnerable People*, edited by Jonathan A. Cook, Owen Cylke, Donald F. Larson, John D. Nach, and Pamela Stedman-Edwards, 14–41. Cheltenham: Edward Elgar.

Pascual, Miguel A., and Javier E. Ciancio. 2007. "Introduced Anadromous Salmonids in Patagonia: Risks, Uses, and a Conservation Paradox." In *Ecological and Genetic Implications of Aquaculture Activities*, edited by Theresa M. Bert, 333–53. Houten: Springer Netherlands.

Penner, Liisa. 2005. *Salmon Fever, River's End: Tragedies on the Lower Columbia River in the 1870s, 1880s, and 1890s: Articles from Astoria Newspapers*. Portland, OR: Frank Amato.

Petersen, David. 2007. "The Frontier Spirit in Japan." *AuthorsDen*, December 27, 2007. www.authorsden.com/visit/viewArticle.asp?id=35529.

Phyne, John, and Jorge Mansilla. 2003. "Forging Linkages in the Commodity Chain: The Case of the Chilean Salmon Farming Industry, 1987–2001." *Sociologia Ruralis* 43 (2): 108–27.

Pomeranz, Kenneth, and Steven Topik. 2014. *The World That Trade Created: Society, Culture and the World Economy, 1400 to the Present*. 7th ed. London: Routledge.

Pulvers, Roger. 2015. "Illusions of the Self: The Life and Poetry of Ishikawa Takuboku." *Asia-Pacific Journal: Japan Focus* 13 (15-2): Article ID 4306. https://apjjf.org/Roger-Pulvers/4306.html.

Quinn, Thomas P., Jeramie A. Peterson, Vincent F. Gallucci, William K. Hershberger, and Ernest L. Brannon. 2002. "Artificial Selection and Environmental Change: Countervailing Factors Affecting the Timing of Spawning by Coho and Chinook Salmon." *Transactions of the American Fisheries Society* 131 (4): 591–98.

Quinones, Renato A., Marcelo Fuentes, Rodrigo M. Montes, Doris Soto, and Jorge León-Muñoz. 2019. "Environmental Issues in Chilean Salmon Farming: A Review." *Reviews in Aquaculture* 11 (2): 375–402.

Racel, Masako N. 2011. "Finding Their Place in the World: Meiji Intellectuals and the Japanese Construction of an East-West Binary, 1868–1912." Master's thesis, Georgia State University.

Rand, Peter. 2020. "International Union of Conservation Advisory Mission Shiretoko (Japan), September 23–20, 2019, February 2020 Report." https://whc.unesco.org/en/list/1193/documents/.

Rees, William E. 1992. "Ecological Footprints and Appropriated Carrying Capacity: What Urban Economics Leaves Out." *Environment and Urbanization* 4, no. 2 (October): 121–30.

Ritvo, Harriet. 1987. *The Animal Estate: The English and Other Creatures in the Victorian Age.* Cambridge, MA: Harvard University Press.

Roberts, Luke. 2002. *Mercantilism in a Japanese Domain: The Merchant Origins of Economic Nationalism in 18th-Century Tosa.* Cambridge, UK: Cambridge University Press.

Roche, Gerald, Hiroshi Maruyama, and Åsa Virdi Kroik. 2018. *Indigenous Efflorescence: Beyond Revitalisation in Sapmi and Ainu Mosir.* Canberra: ANU Press.

Rossi, C. M. Riva, Miguel Alberto Pascual, E. Aedo Marchant, N. Basso, J. E. Ciancio, Bárbara Mezga, Daniel Alfredo Fernández, and B. Ernst-Elizalde. 2012. "The Invasion of Patagonia by Chinook Salmon (Oncorhynchus Tshawytscha): Inferences from Mitochondrial DNA Patterns." *Genetica* 140, no. 10 (December): 439–53.

Ruggerone, Gregory T., Beverly A. Agler, and Jennifer L. Nielsen. 2012. "Evidence for Competition at Sea between Norton Sound Chum Salmon and Asian Hatchery Chum Salmon." *Environmental Biology of Fishes* 94, no. 1 (May): 149–63.

Ruggerone, Gregory T., and James R. Irvine. 2018. "Numbers and Biomass of Natural- and Hatchery-Origin Pink Salmon, Chum Salmon, and Sockeye Salmon in the North Pacific Ocean, 1925–2015." *Marine and Coastal Fisheries* 10, no. 2 (April): 152–68.

Russell, Harold S. 2007. *Time to Become Barbarian: The Extraordinary Life of General Horace Capron.* New York: University Press of America.

Saavedra-Rivano, Neantro. 1993. "Chile and Japan: Opening Doors through Trade." In *Japan, the United States, and Latin America: Toward a Trilateral Relationship in the Western Hemisphere*, edited by Barbara Stallings and Gabriel Szekely, 191–209. Baltimore, MD: Johns Hopkins University Press.

Sahashi, Genki. 2021. "Homogenization of the Timing of Chum Salmon Returns in Relation to Hatchery Transplantation." *Environmental Biology of Fishes* 104 (2): 135–42.

Said, Edward. 1978. *Orientalism*. New York: Pantheon Books.

Saitama Kenritsu Hakubutsukan [Saitama Prefectural River Museum]. 1998. "Bankoku hakurankai to sake no jinkō fuka jigyō tokubetsuten michikusa tenbyō" [Special exhibit idle sketches on international exhibitions and the man-made salmon hatcheries industry]. *Kawahaku Dayori* [Kawahaku newsletter], August 1.

Salmon Chile. n.d. "Exports." www.salmonchile.cl/en/exports/.

Sapporo Agricultural College. 1878. *Annual Report*. Hokkaidō Daigaku Tosho Kankōkai.

Sato, Shunpei, and Kentaro Morita. 2019. "Hokkido ni okeru sakeyaseigyo no identeki tokuchō" [Genetic uniqueness of wild chum salmon populations in Hokkaido, Japan]. *Journal of the Ecological Society of Japan* 69 (3): 209–17.

Sato, Shunpei, William D. Templin, Lisa W. Seeb, James E. Seeb, and Shigehiko Urawa. 2014. "Genetic Structure and Diversity of Japanese Chum Salmon Populations Inferred from Single-Nucleotide Polymorphism Markers." *Transactions of the American Fisheries Society* 143 (5): 1231–46.

Satsuka, Shiho. 2015. *Nature in Translation*. Durham, NC: Duke University Press.

Segawa, Takuro. 2005. *Ainu-ekoshisutemu no kōkogaku: Ibunka kōryū to shizen riyō kara mita ainu shakai seiritsushi* [Archaeology of Ainu ecosystems: A history of Ainu social development as seen through cross-cultural trade and use of natural resources]. Hokkaidō Shuppan Kikaku Sentaa [Hokkaido publishing planning center].

——. 2007. *Ainu no rekishi: Umi to takara no nomado* [History of the Ainu: Nomads of the ocean and treasures]. Tokyo: Kôdansha.

Seidensticker, Edward G. 1951. "Japanese Fisheries Reform: A Case Study." *Far Eastern Survey* 20, no. 18 (October): 185–88.

Seufert, Francis. 1980. *Wheels of Fortune*. Portland: Oregon Historical Society.

Shimizu, Ikutaro. 2005. "Economic Factors Effecting Salmon Fisheries in Japan." *Bulletin of the National Salmon Resources Center* 7 (March): 105–15.

Shimura, Shigeru, Eduardo Cardenas, and Aliaky Nagasawa. 1986. "Introduction into Aysen Chile of Pacific Salmon No 17." Servicio Nacional de Pesca, Ministerio de Economía Fomento y Reconstrucción, República de Chile, and Japan International Cooperation Agency.

Shirai, Nobuaki. 2010. *Hokkaidō kaitakusha seishin to kirisutokyō* [Christianity and the spirit of Hokkaido settlers]. Sapporo: Hokkaidō Daigaku Shuppankai [Hokkaido university publishing].

Shreve, Bradley G. 2009. "'From Time Immemorial': The Fish-In Movement and the Rise of Intertribal Activism." *Pacific Historical Review* 78, no. 3 (August): 403–34.

Siddle, Richard. 1996. *Race, Resistance and the Ainu of Japan*. New York: Routledge.

———. 1999. "Ainu History: An Overview." In *Ainu: Spirit of a Northern People*, edited by William W. Fitzhugh and Chisato O. Dubreuil, 67–73. Washington, DC: Arctic Studies Center, National Museum of Natural History, Smithsonian Institution, in association with University of Washington Press.

Skabelund, Aaron. 2011. *Empire of Dogs*. Ithaca, NY: Cornell University Press.

Stanlaw, James. 1992. "For Beautiful Human Life: The Use of English in Japan." In *Re-made in Japan: Everyday Life and Consumer Taste in a Changing Society*, edited by Joseph J. Tobin, 58–76. New Haven, CT: Yale University Press.

———. 2004. *Japanese English: Language and Culture Contact*. Hong Kong: Hong Kong University Press.

Statistics Japan. 2020. *Statistical Handbook of Japan*. Tokyo: Ministry of Internal Affairs and Communications Japan, Statistics Bureau. www.stat .go.jp/english/data/handbook/index.html.

Steele, M. William. 2007. "Casting Shadows on Japan's Enlightenment: Sada Kaiseki's Attack on Lamps." International Christian University Publications 3-A. *Asian Cultural Studies Special Issue* 16:57–73.

Stoler, Ann Laura. 2001. "Tense and Tender Ties: The Politics of Comparison in North American History and (Post) Colonial Studies." *Journal of American History* 88, no. 3 (December 1): 829–65.

———, ed. 2006. *Haunted by Empire: Geographies of Intimacy in North American History*. Durham, NC: Duke University Press.

Sunset Magazine. 1921. "The Pulse of the West: Soaking the Sockeye in Puget Sound." *Sunset Magazine* 47, no. 6 (December): 15.

Swanson, Heather Anne. 2017. "Methods for Multispecies Anthropology: Thinking with Salmon Otoliths and Scales." *Social Analysis* 61, no. 2 (June): 81–99.

———. 2018. "Landscapes, by Comparison." In *The World Multiple: The Quotidian Politics of Knowing and Generating Entangled Worlds*, edited by Keiichi Omura, Grant Jun Otsuki, Shiho Satsuka, and Atsuro Morita, 105–22. New York: Routledge.

———. 2019. "The Entrapment of Trap Design: Materiality, Political Economy and the Shifting Worlds of Fixed Gear Fishing Equipment." *Journal of Material Culture* 24, no. 4 (March): 401–20.

Swanson, Heather Anne, Nils Bubandt, and Anna Tsing. 2015. "Less than One but More than Many: Anthropocene as Science Fiction and Scholarship-in-the-Making." *Environment and Society* 6, no. 1 (September): 149–66.

Takagi, Shinji. 1995. *From Recipient to Donor: Japan's Official Aid Flows, 1945 to 1990 and Beyond*. Princeton, NJ: Princeton University Press.

Takahashi, Mitsuhiko. 2009. "Overview of the Structure and the Challenges of Japanese Wildlife Law and Policy." *Biological Conservation* 142 (9): 1958–64.

Takeuchi, Yoshimi. 2005. *What Is Modernity? Writings of Takeuchi Yoshimi*. New York: Columbia University Press.

Tama, Shinnosuke. 2012. "Hokkaido Farming Methods in Manchukuo in 1940s." *Regional Science Research, University of Tokushima* 2:30–41.

Tanabe, Yoko. 2019. "A Historical Perspective of Indigenous Education Policy in Japan: The Case of Ainu Schools." In *Sámi Educational History in a Comparative International Perspective*, edited by Otso Kortekangas, Pigga Keskitalo, Jukka Nyyssönen, Andrej Kotljarchuk, Merja Paksuniemi, and David Sjögren, 207–24. Cham: Palgrave Macmillan.

Taylor, Joseph E. 1999. *Making Salmon: An Environmental History of the Northwest Fisheries Crisis*. Seattle: University of Washington Press.

Tetlow, Roger T., and Graham J. Barbey. 1990. *Barbey: The Story of a Pioneer Columbia River Salmon Packer*. Portland, OR: Binford & Mort.

Tillotson, Michael, Heidy Barnett, Mary Bhuthimethee, Michele Koehler, and Thomas Quinn. 2019. "Artificial Selection on Reproductive Timing in Hatchery Salmon Drives a Phenological Shift and Potential Maladaptation to Climate Change." *Evolutionary Applications* 12 (7): 1344–59.

Togawa, Tsuguo. 1995. "The Japanese View of Russia before and after the Meiji Restoration." In *A Hidden Fire: Russian and Japanese Cultural Encounters, 1868–1926*, edited by J. Thomas Rimer, 214–27. Redwood City, CA: Stanford University Press.

Townsend, Ralph Edwin, R. Townsend, Ross Shotton, and Hirotsugu Uchida, eds. 2008. "Case Studies in Fisheries Self-Governance." *FOA Fisheries Technical Paper*, No. 504. Rome: Food & Agriculture Org.

Treat, Upham S. 1878. "Hokkaidō gyogyō hōkokusho" [Report on Hokkaido fishing]. Upham S. Treat 009. Kaitakushi gaikokujin kankei shokan moku-roku [Catalog of Hokkaido Development Commission correspondence with foreigners], Hokkaido University Library Northern Studies Collection.

Tsing, Anna Lowenhaupt. 1993. *In the Realm of the Diamond Queen: Marginality in an Out-of-the-Way Place*. Princeton, NJ: Princeton University Press.

———. 2005. *Friction: An Ethnography of Global Connection*. Princeton, NJ: Princeton University Press.

———. 2015. *The Mushroom at the End of the World: On the Possibility of Life in Capitalist Ruins*. Princeton, NJ: Princeton University Press.

Tsing, Anna Lowenhaupt, Nils Bubandt, Elaine Gan, and Heather Anne Swanson, eds. 2017. *Arts of Living on a Damaged Planet: Ghosts and Monsters of the Anthropocene*. Minneapolis: University of Minnesota Press.

Tsing, Anna Lowenhaupt, Andrew S. Mathews, and Nils Bubandt. 2019. "Patchy Anthropocene: Landscape Structure, Multispecies History, and the Retooling of Anthropology: An Introduction to Supplement 20." *Current Anthropology* 60, no. S20 (August): S186–S197.

Tyrrell, Ian. 1991. "American Exceptionalism in an Age of International History." *American Historical Review* 96, no. 4 (October): 1031–55.

———. 1999. *True Gardens of the Gods: Californian-Australian Environmental Reform, 1860–1930*. Berkeley: University of California Press.

Ukaji, Shizue. 2018. "The Racing of Ainu Hearts: Our Wish for One Salmon River." Translated by Miku Maeda. In *Indigenous Efflorescence: Beyond Revitalisation in Sapmi and Ainu Mosir*, edited by Gerald Roche, Hiroshi Maruyama, and Asa Virdi Kroik, 79–83. Acton: Australian National University Press.

Urrutia, Claudio Rosales. 2007. "Los sectores archipielágicos del Chile sur austral: Una historia que aun está lejos de concluir a través de la industria del salmon." *XII Jornadas Nacionales de Historia Regional de Chile, Universidad de La Serena* 12a:455–78.

US Congress. 1862. "Morrill Act, Chap. CXXX.—An Act Donating Public Lands to the Several States and Territories Which May Provide Colleges for the Benefit of Agriculture and Mechanic Arts. SEC. 4." www.ourdocuments.gov/doc.php?doc=33&page=transcript.

US Fish & Wildlife Service. n.d. "Mass Marking Coded Wire Tagging, and Passive Integrated Transponder Tagging of Salmon and Steelhead." www.fws.gov/wafwo/pdf/mmbrochure.pdf.

Uzawa, Kanako. 2018. "Everyday Acts of Resurgence and Diasporic Indigeneity among the Ainu of Tokyo." In *Indigenous Efflorescence: Beyond Revitalisation in Sapmi and Ainu Mosir*, edited by Gerald Roche, Hiroshi Maruyama, and Åsa Virdi Kroik, 179–203. Canberra: ANU Press.

———. 2020. "'Crafting Our Future Together': Urban Diasporic Indigeneity from an Ainu Perspective in Japan." PhD diss., UiT The Arctic University of Norway.

Uzawa, Kanako, and Mark K. Watson. 2020. "Urespa ('Growing Together'): The Remaking of Ainu-Wajin Relations in Japan through an Innovative Social Venture." *Asian Anthropology* 19, no. 1 (January): 53–71.

Vanstone, James. 1993. "The Ainu Group at the Louisiana Purchase Exposition, 1904." *Arctic Anthropology* 30, no. 2 (January 1): 77–91.

Viveiros de Castro, Eduardo. 2004. "Perspectival Anthropology and the Method of Controlled Equivocation." *Tipití: Journal of the Society for the Anthropology of Lowland South America* 2, no. 1 (January): 3–22.

Wada, Eita. 1994. *Sake to kujira to nihonjin: Sekizawa Akiyo no shôgai* [Salmon, whales and Japanese: The life of Sekizawa Akiyo]. Tokyo: Naruyamadô Shoten.

Walker, Brett. 1999. "Foreign Contagions, Ainu Medical Culture, and Conquest." In *Ainu: Spirit of a Northern People*, edited by William W. Fitzhugh and Chisato O. Dubreuil, 102–7. Washington, DC: Arctic Studies Center, National Museum of Natural History, Smithsonian Institution, in association with University of Washington Press.

———. 2001. *The Conquest of Ainu Lands: Ecology and Culture in Japanese Expansion, 1590–1800*. Berkeley: University of California Press.

———. 2004. "Meiji Modernization, Scientific Agriculture, and the Destruction of Japan's Hokkaido Wolf." *Environmental History* 9 (2): 248–74.

———. 2005. *The Lost Wolves of Japan*. Seattle: University of Washington Press.

Walker, Peter. 2005. "Political Ecology: Where Is the Ecology?" *Progress in Human Geography* 29, no. 1 (February): 73–82.

Wallerstein, Immanuel. 2004. *World-Systems Analysis: An Introduction*. Durham, NC: Duke University Press.

Walsh, Hall. 1877. "Kaitakushisei sake kanzume shishoku hinpyō" [Evaluation of a taste-test of Hokkado development commission canned salmon]. 0C03883000000000. Kaitakushi gaikokujin kankei shokan mokuroku [Catalog of Hokkaido Development Commission correspondence with foreigners], Hokkaido University Library Northern Studies Collection.

Waples, Robin. 1991. "Pacific Salmon, Oncorhynchus spp., and the Definition of 'Species' under the Endangered Species Act." *Marine Fisheries Review* 53 (3): 11–22.

Waples, Robin, George R. Pess, and Tim Beechie. 2008. "Evolutionary History of Pacific Salmon in Dynamic Environments." *Evolutionary Applications* 1, no. 2 (May): 189–206.

Watson, Mark K. 2014. *Japan's Ainu Minority in Tokyo: Diasporic Indigeneity and Urban Politics*. London: Routledge.

Watsuji, Tetsurō. (1935) 1988. *Climate and Culture: A Philosophical Study*. New York: Greenwood.

Welch, Laine. 2020. "Hatchery Hauls: Alaska's Statewide Salmon Catch Gets $118 Million Boost." *National Fisherman*. www.nationalfisherman.com /alaska/hatchery-hauls-alaskas-statewide-salmon-catch-gets-118-million -boost.

White, Cliff. 2016. "Mitsubishi Merges Salmones Humboldt into Cermaq Chile." *SeafoodSource*. www.seafoodsource.com/news/aquaculture /mitsubishi-merges-salmones-humboldt-into-cermaq-chile.

White, Richard. 1991. *The Middle Ground: Indians, Empires, and Republics in the Great Lakes Region, 1650–1815*. Cambridge, UK: Cambridge University Press.

———. 1995. *The Organic Machine*. New York: Hill and Wang.

Wild Fish Conservancy Northwest. 2021. "Washington Takes Historic Step to Legalize Fish Traps for Sustainable Commercial Fishing on the Columbia River." https://wildfishconservancy.org/washington-takes-historic-step-to -legalize-fish-traps-for-sustainable-commercial-fishing-on-the-columbia -river.

Willcock, Hiroko. 2000. "Traditional Learning, Western Thought, and the Sapporo Agricultural College: A Case Study of Acculturation in Early Meiji Japan." *Modern Asian Studies* 34 (4): 977–1017.

Winn, Peter, and Cristobal Kay. 1974. "Agrarian Reform and Rural Revolution in Allende's Chile." *Journal of Latin American Studies* 6, no. 1 (May 1): 135–59.

Woody, E., E. C. Wolf, and S. Zuckerman. 2003. *Salmon Nation: People, Fish, and Our Common Home*. Corvallis: Oregon State University Press.

Worster, Donald. 1985. *Rivers of Empire: Water, Aridity, and the Growth of the American West*. New York: Pantheon.

Yabe, K. 1993. "Wetlands of Hokkaido." In *Biodiversity and Ecology in the Northernmost Japan*, edited by Seigo Higashi, Akira Osawa, and Kana Kanagawa, 38–49. Sapporo: Hokkaido University Press.

Yaguchi, Yujin. 2002. "American Objects, Japanese Memory: 'American' Landscape and Local Identity in Sapporo, Japan." *Winterthur Portfolio* 37, no. 2/3 (June 1): 93–121.

Yamamoto, Tadashi. 1995. "Development of a Community-Based Fishery Management System in Japan." *Marine Resource Economics* 10, no. 1 (Spring): 21–34.

Yamaura, Kiyoshi, and Hiroshi Ushiro. 1999. "Prehistoric Hokkaido and Ainu Origins." In *Ainu: Spirit of a Northern People*, edited by William W. Fitzhugh and Chisato O. Dubreuil, 39–46. Washington, DC: Arctic Studies Center, National Museum of Natural History, Smithsonian Institution, in association with University of Washington Press.

Yasuda, Sadamori. 1876. "Ainufū shiozake (kan'en hiki) shintei (hikae)/ shōbankan Yasuda Sadamori" [Offering of Ainu-style salted salmon (memo)/Judge Yasuda Sadamori]. Amerika kōshikan 032. Kaitakushi gaikokujin kankei shokan mokuroku [Catalog of Hokkaido Development Commission correspondence with foreigners], Hokkaido University Library Northern Studies Collection.

———. 1878. "Kaitakushi sake kanzume hitobako sōfu nitsuki hinpyō irai narabi hanro mikomi shōkai (eibun hikae)/Kendai shokikan Yasuda Sadamori (Tokyo)" [Request for evaluation of a box of canned salmon from the Hokkaido Development Commission and an inquiry for prospects of sales routes [English memo]/Secretary Yasuda Sadamori (Tokyo)]. Maruseiyu Nihon Ryōjikan [Japanese Consulate of Marseille] 002. Kaitakushi gaikokujin kankei shokan mokuroku [Catalog of Hokkaido Development Commission correspondence with foreigners], Hokkaido University Library Northern Studies Collection.

INDEX

agricultural development: Ainu assimilation, Japaneseness, and, 173–74; Kaitakushi and, 42–47; in Latin America, 72–73; livestock, 44–45; Model Barn, 48, 49*fig.*; river concretization and, 145

Ainu Association of Hokkaido, 180, 181, 185

Ainu Cultural Promotion Act (1997), 178, 215n24

Ainu peoples: agricultural history of, 173, 203n17; Ainuness and resurgence, 176–79; assimilation policies, 45–46, 166–67, 172–76; bans against fishing and hunting by, 46, 174–75; *basho ukeoi* direct-labor system, 170, 172, 173; civilizing narratives and, 39; "cowboys and Indians" trope and, 166–67; "cultural promotion" salmon harvest permits, 178–79, 185; history of, 168–70; Indigenous rights and, 177–79, 182–87; land seizures, 46; Monbetsu Ainu, 179–88; Nibutani Dam and, 215n28; numbers of, 170, 214n19; Raporo Ainu, 187; salmon processing history, 170–71; schools, segregated,

46, 203n18; uplift narratives, 173–74; waste disposal site struggle, 182–85

Alaska: canning, 204n26; hatcheries, 211n11; Indigenous activism, 186; price declines, 208n1; set-net ban, 210n9; wild salmon and, 143

Allende, Salvador, 76

American model, 42–46

American occupation policies, 122, 124–25, 209n6

American West: comparison, noncomparability, and, 195–97; "cowboys and Indians" trope, 166–67; Meiji modernization and comparison of Hokkaido to, 39–41, 63–64; as "modern," 13; New Western History, 10–11. *See also* Columbia River basin

Anderson, Benedict, 26, 27–28

Anthropocene, 5–6

Arnold, David, 10–11

Asahikawa project, 144–49, 164

assimilation policies, 45–46, 166–67, 172–76

astaxanthin, 207n11

Atlantic salmon, 94, 102

auctions, 138

color of fish and skin, 95, 102, 207n4, 207n11

Columbia River basin, xviimap, 22; canning, 53, 55fig., 55–57, 60–61, 196; incomparability and, 196–97; Indigenous activism, 185–86; land-use practices, 211n12; material conditions for wild salmon, 157–59; stock decline, 154; wild salmon and hatchery fish, 153–56

Columbia River Inter-Tribal Fish Commission, 186

commodified relations to fish in Kitahama, 135–40

commodity chains, 64, 67, 73–74, 110–11. See also JICA-Chile salmon project

comparison and comparative practices: American West, exceptionalism, and noncomparability, 194–97; biographies and emergence of, 67; Chilean salmon shadow effects and, 109–14; China and, 29; commodity chains and, 110; concretized histories of, 164–65; cross-cultural, in anthropology, 24–28; as ethnographic object, 20–21; Japan-West binary and, 28–36; materiality of, 16–19; nation-states, material world-making, and, 11–16; rice people, bread people, and salmon people, 176, 187; "specters of comparison," 193, 196–97; temporal, 148, 162; way of thinking, fisher cosmopolitan identities, and, 117–18; as world-making practices, 17. See also Ainu peoples; Chilean salmon industry; Columbia River basin; hatcheries; JICA-Chile salmon project;

landscape comparison; wild salmon management

concretization of rivers, 62, 70, 145–46, 163–65

conservation projects, 110, 113–14

Convention for the Conservation of Anadromous Stocks in the North Pacific Ocean (1993), 69

cooperatives: Ainu and, 180; hatcheries and, 112; Monbetsu River waste dump and, 183; organizational structure in Kitahama, 129–30; price declines and, 109–10; self-governance principles and, 125; watershed conservation and, 163. See also Kitahama fishers

Cousiño, Louis, 75

"cowboys and Indians" trope, 166–67

Cronon, William, 10

cross-cultural comparison in anthropology, 24–28

Cutter, John, 59

Dai Nippon Suisan Kai, 70

Daisetsuzan National Park, 145, 146

data and number crunching, 131

Dauvergne, Peter, 111

Dawes General Allotment Act (US, 1887), 46, 203n16

deer, Ezo, 176

development aid, 71, 74, 82. See also JICA-Chile salmon project

dikes, 146

Dower, John, 33–34

dried salmon, 137, 169–71

Dun, Edwin, 44–45

ecological impacts of farmed salmon, 107–8

eggs, transport of, 85

place name and Ezo vs., 38–39; as tabula rasa, 37–38

Hokkaido Colonization Commission (Kaitakushi), 42–47, 51–58, 174, 194–95

Hokkaido Federation of Fisheries Cooperatives, 109–10, 142

Hokkaido Industrial Pollution Examination Panel, 185

Hokkaido University, 63

Hokusui Kyōkai (fisheries society), 61

Honshu compared to Hokkaido, 41–42, 63–64

Honshu salmon, 23, 201n1(ch.1)

Hume brothers, 53

Indigenous rights, 177–79, 182–87. *See also* Ainu peoples

Inuit, 177

Ishikari, Hokkaido, 3–4

Ishikari River, 54, 144–45

Ishikawa, Takuboku, 37–38

Ito, Kazutaka, 60–62

Iwakura Mission, 42

Japanese linguistics, 17–18

Japan International Cooperation Agency (JICA), 205n3. *See also* JICA-Chile salmon project

JICA-Chile salmon project: background, 65–71, 82–83; brood stock program, 86–87; brown trout, predation by, 86; Chilean desires, 75–79; comparison and, 67–68; establishment of Coyhaique hatchery, 83–84; failure of fish to return, 86–88; as failure or success, 89–90, 92, 104–6; fish diet, 93; influence on Chilean salmon industry, 91–96; Japanese desires,

71–74; Nagasawa-san and, 67–69, 79–90; resource nationalism and, 69; Rio Ultima Esperanza, adult fish found in, 88–89; season, issues of, 86–87; transportation of eggs, 85–86

Kaitakushi (Hokkaido Colonization Commission), 42–47, 51–58, 174, 194–95

kangaekata (way of thinking), 117

katakana, 18

katei ryōri (Japanese homestyle cuisine), 102

Kawakami Takiya, 205n1

Kayano, Shigeru, 46, 175

keiji (sexually immature salmon), 139–40

kenkō ni ii (good for health), 103

Kitahama fishers: Chinese processors and, 132–33; collective organizational structure, 129–30; commodified relations to fish, 135–40; data and rationalization, 130–31; fishing rights, hereditary, 123–27, 209n7; fractional shares system, 131–32; increase in salmon, price decline, and, 128; marginality of Kitahama, 118–21; modern selfhood, cosmopolitan identities, and, 117–18, 124, 133–35, 136; MSC eco-label and, 141–42, 159–60; outmigration and return, generational, 121–24; self-management, 116–17; set-net traps, 125–27, 126*fig.*, 134*fig.*; stereotypes of fishermen and, 115–16, 120–21

Komaba Agricultural College, 42

Kuril islands, xvii*map*, 57, 66, 69, 201n3(ch.3), 201n5, 213n5

Kuroda, Kiyotaka, 42–43, 47, 48, 54

CULTURE, PLACE, AND NATURE
Studies in Anthropology and Environment

www.ingramcontent.com/pod-product-compliance
Lightning Source LLC
Chambersburg PA
CBHW031415270326
41929CB00010BA/1468